CLIMATE CHANGE SCIENCE

COLUMBIA UNIVERSITY EARTH INSTITUTE
SUSTAINABILITY PRIMERS

COLUMBIA UNIVERSITY EARTH INSTITUTE
SUSTAINABILITY PRIMERS

The Earth Institute (EI) at Columbia University is dedicated to innovative research and education to support the emerging field of sustainability. The Columbia University Earth Institute Sustainability Primers series, published in collaboration with Columbia University Press, offers short, solutions-oriented texts for teachers and professionals that open up enlightened conversations and inform important policy debates about how to use natural science, social science, resource management, and economics to solve some of our planet's most pressing concerns, from climate change to food security. The EI Primers are brief and provocative, intended to inform and inspire a new, more sustainable generation.

Renewable Energy: A Primer for the Twenty-First Century, Bruce Usher

CLIMATE CHANGE SCIENCE

A PRIMER FOR SUSTAINABLE DEVELOPMENT

JOHN C. MUTTER

Columbia University Press　*New York*

Columbia University Press
Publishers Since 1893
New York   Chichester, West Sussex
cup.columbia.edu
Copyright © 2020 John C. Mutter
All rights reserved

Cataloging-in-Publication Data available from the
Library of Congress.

ISBN 978-0-231-19222-4 (cloth)
ISBN 978-0-231-19223-1 (paperback)
ISBN 978-0-231-54972-1 (ebook)

LCCN 2019046850

Cover design: Julia Kushnirsky
Cover photograph: Tas van Ommen

CONTENTS

ACKNOWLEDGMENTS

I MUST first acknowledge my dear friend and colleague Mark Cane, the G. Unger Vetlesen Professor in Columbia's Department of Earth and Environmental Sciences. In one sense, Mark had no role in this primer—he did not review or critique it in any way, nor did he contribute a single word of text or an illustration. But in another way, Mark is wholly responsible for the book. I teach courses in science to undergraduates and graduate students who are primarily interested in something else, such as microeconomics. When I first started teaching at a graduate level, I asked Mark to give a few lectures on climate science because I did not have the confidence to do it myself. Mark is so clear, and his illustrations so enlightening, that I absorbed his lectures easily and wholly, learning as much as any student. When I subsequently began teaching climate science myself, Mark's slide materials and my memory of his presentations became the foundation for my lectures. The structure of this primer follows the structure of Mark's lectures.

My deepest thanks to Claire Palandri, a second-year PhD student (at the time) in the sustainable development program I direct. She read and commented on an earlier draft of the primer. Almost every paragraph and most of the figures had comments from Claire that were extremely helpful. I am not exaggerating when I say that I acted on every comment and that every comment led to an improvement.

Then I must thank Hilary Osborn, a student in the sustainability management program at Columbia, who undertook the monumental task of turning colored figures into grayscale images, searched for suitable figures for grayscale conversion, found sources for figures whose origin I had forgotten, and completed many other tasks. Hilary also read and critiqued the manuscript and provided helpful suggestions. She put in many hours on the project, cheerfully taking on some of the most tedious tasks. Had I needed to do that myself, this book would not exist.

I also am indebted to Gauthami Ram Mohan, who took over from Hilary once she completed her studies at Columbia and moved on to the real world of gainful employment. Gauthami joined the project as an engineering master's student who volunteered to help so she might learn about sustainable development issues. She, too, cheerfully took on some fairly tedious tasks with skill and dedication.

Finally, I express my deep appreciation to Jeffrey T. Kiehl for his very thorough and immensely helpful review of an earlier version of the manuscript, which he returned with extensive comments that I acted on in every case. He not only pointed out my errors, of which there were no small number, but also suggested how to correct them. It is a rare reviewer who will do that, and I am deeply grateful.

INTRODUCTION

Strategy and Outline of the Primer

T HIS book provides the scientific basis required to assess what changes to the climate might plausibly come about in the future and what might cause these changes to occur. It provides the basics of physical climatology set out with the objective of developing an understanding of how human activity might lead to changes in climate and what form those changes might take. As such, it does not attempt to cover all aspects of climatology. One central theme is to elaborate the origin of the inherent and irreducible uncertainty in predictions of possible climate futures.

This text emphasizes physical aspects of the climate system—winds, temperature, ocean currents, atmospheric circulation, and rainfall—both in building an understanding of the mechanics of the climate system and in describing drivers of change to the system and potential consequences, such as sea level rise and hurricane dynamics. Aspects of the climate system and climate change that depend upon and may influence ecological systems are dealt with in other books in the Columbia University Earth Institute Sustainablity Primers series.

Most natural science fields, including climate science, make gains through a process of successive approximations. The first approximation describes the most fundamental aspect of the system under study. For example, as a first approximation, we know that atoms are made of neutrons, protons, and electrons. Although we now know that atoms are much more complex than this, with scores of subatomic particles, that knowledge has not made the initial description incorrect—there are still neutrons, protons, and electrons. However, newer and progressively more refined approximations can be made that lead to a more complete description of the atom. No physicist would suggest, even with verification of the *Higgs boson*, that atoms are now fully described. In the same way, no climate scientist would suggest that the climate system has been fully described and understood. It is, however, known at a level sufficient to say how human activity can influence climate.

Bertrand Russell expressed this idea succinctly: "Science does not aim at establishing immutable truths and eternal dogmas; its aim is to approach the truth by successive approximations, without claiming that at any stage final and complete accuracy has been achieved."[1] The structure of this text follows Russell's theme of successive approximations in describing our understanding of the climate system. In taking this approach, I also follow a narrative arc that takes us from some of the earliest descriptions of the climate system to the present day, where much of the focus is on climate prediction. I also discuss some of the consequences of climate change in the context of sustainable development.

In the first approximation, we imagine Earth to be essentially static. There are no oceans and no vegetation, and the only feature of the atmosphere is clouds. From this starting point, I develop a theory that explains why Earth's average temperature is around 16°C. This leads directly to the greenhouse effect, using basic elements of well-established atomic physics, and to the inference that human activity can indeed alter climate through modification of the atmosphere's chemical composition.

The second level of approximation acknowledges that Earth is a spheroid rotating on its axis, and that the temperature varies with location on the planet. From these facts, I develop an understanding of wind patterns, atmospheric and oceanic circulation, plus rainfall patterns and their variations. Developing this second-order approximation in no way invalidates conclusions drawn from the first level of approximation, but it adds features and details of the climate system not available in the first approximation. The second approximation draws on classical Newtonian mechanics.

The next level of refinement, the third approximation, examines several dynamic features of climate, including the glacial/interglacial cycles, the El Niño Southern Oscillation, and other features of Earth's natural dynamics. Here we need to include some aspects of the behavior of the ocean and its coupled interaction with the atmosphere. This chapter includes a discussion of the consequences of oscillating climate states on malaria prevalence and crop production in poor countries. A discussion of the issues around prediction of dynamic systems is also included.

Through these steps, a sufficient understanding of the climate system is built so we can analyze the critical question of Earth's future climate through use of climate model projections. In particular, I examine the reasons for uncertainties when predicting the climate's distant future.

I conclude with three examples of the consequences of climate warming that illustrate how particular aspects of Earth's system might be modified under climate change (through sea level rise and tropical cyclones) together with some comments on future climate variability and committed warming, which is the warming to be expected even were the concentration of greenhouse gases to stabilize.

• • •

The text is derived from class notes prepared by the author for a course titled Science for Sustainable Development, cotaught with Professor Ruth DeFries to undergraduates at Columbia University. The class notes have proven useful for some master's and PhD students as well.

Material contained in boxes is more technical in nature and is parenthetical in form; it need not be read to understand the main body of the text. It is included for completeness for readers who have more preparation in the physical sciences and who may wish to have a greater level of detail on certain subjects.

A glossary contains terms that may be unfamiliar to readers new to the field. These terms, when first introduced in the text, appear in italics. Illustrations are taken from the published literature in most cases, and they generally began as color images that have been transformed to grayscale images, with varying levels of success. References to the original color images are provided in the captions to the figures.

CLIMATE CHANGE
SCIENCE

1

WHY DOES EARTH HAVE
THE CLIMATE IT DOES?

THE LARGE-SCALE PATTERN OF EARTH'S
CLIMATE AND VEGETATION

Developing an understanding of Earth's physical climate begins with a careful description of the main observed features of the system at the broadest scale. Figure 1.1 shows the major climate zones of the world, and figure 1.2 is a similar global map showing major *vegetation* zones. In Bertrand Russell's schema, this description could be thought of as the first approximation analysis of climate. These large-scale persistent features of climate must be accounted for before any further refinement can be considered.

Each hemisphere has three climate zones: tropical, temperate, and polar. Each of the three zones occupies about 30° of latitude. These zones can, of course, be divided into many subzones, but I begin with this very coarse depiction.

There is a clear, broad, spatial correspondence between these two maps, from which it can be concluded that climate has a strong influence on vegetation regimes. This should come as no surprise. Figure 1.3 is a common way to display the dependence of major vegetation types on temperature and precipitation.

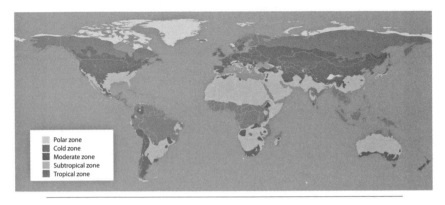

FIGURE 1.1 Simplified climate zones.

Source: Map based on data from M. C. Peel, B. L. Finlayson, and T. A. McMahon, "Updated World Map of the Köppen-Geiger Climate Classification," *Hydrology and Earth System Sciences* 11 (2007), https://doi.org/10.5194/hess-11-1633-2007. Courtesy of Wikimedia Commons.

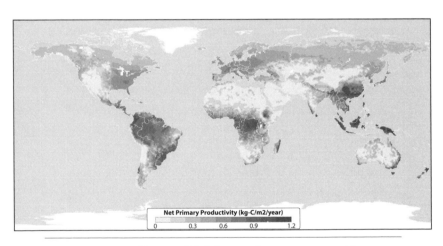

FIGURE 1.2 Pattern of global vegetation types expressed in net primary production (NPP).

Source: "Net Primary Productivity," in *Atlas of the Biosphere*, Center for Sustainability and the Global Environment, University of Wisconsin-Madison, https://nelson.wisc.edu/sage/data-and-models/atlas/maps.php?datasetid=37&includerelatedlinks=1&dataset=37.

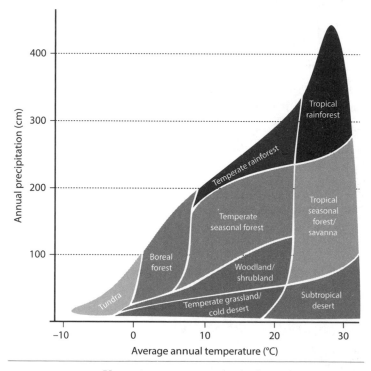

FIGURE 1.3 Vegetation types respond to both temperature and precipitation.

Source: Matthew R. Fisher, ed., *Environmental Biology*, rev. ed. (n.p.: OpenOregon Educational Resources, n.d.), 84, fig. 2, https://openoregon.pressbooks.pub /envirobiology/chapter/3-3-terrestrial-biomes/.

It is referred to as the *Whittaker biome diagram* after its originator, Robert Whittaker, who proposed the scheme in 1975. This diagram can be thought of as a "mapping function" that enables climate zones, described in temperature and precipitation, to be related to vegetation zones, described in species of flora, known as *biomes*.

It is striking that the strongest variation in climate and vegetation type occurs in a north-south (meridional) direction in most parts of the world rather than east to west (see figures 1.1 and 1.2). Tropical regions, where major rain forests are located, occupy the equatorial belt in Africa, South America, and the Indonesian regions. Moving away from the wet tropics in both hemispheres are dry desert regions centered at about 30° north and south. Next are the temperate regions, which include much of eastern North America and Europe. At higher latitudes, both polar regions are frigid and also very dry. The Palmer Dry Plains of Antarctica are one of the driest places on Earth. The pattern is not perfectly symmetrical across the northern and southern hemispheres, but it is almost so. There is much less landmass in the high latitudes of the southern hemisphere, so some of the climate zones observed in the northern hemisphere are not fully represented in the southern hemisphere. The Antarctic is also much colder and dryer than the Arctic, and that has a significant influence on southern hemisphere climate. In large parts of Antarctica, the temperature never rises above freezing, but this is not the case in the Arctic. The two poles are geographical opposites—the Arctic is an ocean surrounded by landmasses, whereas Antarctica is a landmass surrounded by ocean.

Exceptions to the dominant north-south pattern occur in several places. In North America, temperate and dry regions occur at the same latitude with a fairly distinct midcontinent divide, establishing a distinct east-west pattern with dryer regions to the west (figure 1.4). There are also examples in southern Latin America and elsewhere. This orientation is influenced by the effects of ocean currents and associated wind patterns.

The great deserts of Asia are also shifted further toward the pole compared to those of northern Africa, but for a different reason. The Himalayan/Tibetan plateau region influences this shift. Sometimes referred to as the Third Pole, plateau temperatures never reach the low values of true polar regions, especially

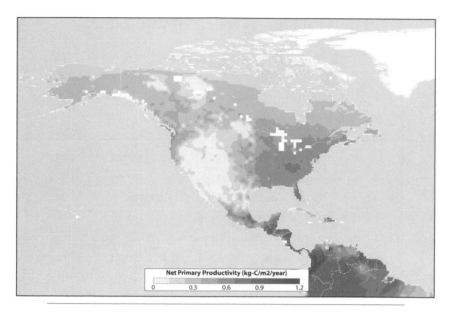

FIGURE 1.4 Changes in vegetation east to west across the United States
and North America.

Source: "Net Primary Productivity," in *Atlas of the Biosphere*, Center for Sustainability
and the Global Environment, University of Wisconsin-Madison,
https://nelson.wisc.edu/sage/data-and-models/atlas/maps.php?
datasetid=37&includerelatedlinks=1&dataset=37.

Antarctica. Summer temperatures in Lhasa, the capital of Tibet,
are above 20°C. The polar climate of the plateau is largely due to
its high elevation, which is not the case for true polar regions.

When we use the word *climate*, we typically mean much more
than temperature. There are both dry and wet tropics, for instance,
and the polar regions are arid and can be thought of as deserts. As
noted, the central region of Antarctica is both the coldest and one
of the driest places on Earth. Parts of the Pacific Ocean could be
described as deserts if rainfall were the only criterion. In addition,
whether vegetation flourishes or not depends on many other factors, including soil type and amount of sunlight.

THE EARTH'S TEMPERATURE FIELD

Figure 1.5 shows Earth's temperature field based on National Aeronautics and Space Administration (NASA) satellite observations. The exact temperature referred to in climate studies must be specified, particularly in climate predictions, and this is the temperature of the air immediately above the surface of Earth regardless of the elevation of the land surface (sometimes referred as the *near-surface temperature*). It is not the temperature of the ground itself; the ground may be much warmer. No correction is made for elevation, so in mountainous regions the temperature is taken on the mountains themselves, and in lowland desert plains the temperature is taken at the low-elevation setting. The image in figure 1.5 shows temperature sensed by satellite;

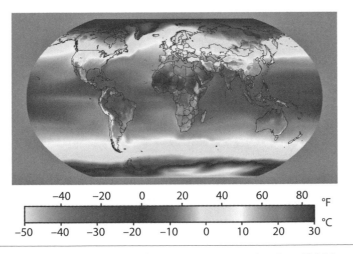

FIGURE 1.5 Global near surface air temperatures, taken from NASA observations. The grayscale makes the regions of high temperature in the low latitudes appear very similar to the very cold polar regions. What is highlighted is the temperate region between the two that appears in a very light shade.

Source: Courtesy of Robert A. Rohde/Wikimedia Commons.

it is not constructed from surface measurements made by ground-based instruments.

Air temperature drops rapidly with elevation above the surface; the average temperature at 30,000 feet—the height at which modern passenger aircraft fly—is around -50°C. Air temperature must be specified at the height within the air column being measured. The near-surface temperature is used because this is the temperature experienced by most plants and animals, including humans (exceptions include the tops of very tall trees and high-flying birds). To the extent that vegetation patterns are a response to temperature, it is this surface air temperature that the vegetation experiences. This is the temperature referred to in weather forecasts, as well as the temperature predicted by global climate models.

In the next section, I discuss atmospheric pressure, which is almost always described as the *sea level pressure* (SLP). If a pressure reading is made by an instrument located on a surface above or below sea level, an adjustment is made to provide the pressure that would have been recorded at sea level.

The temperature scale is shown in degrees Celsius (°C) and in degrees Fahrenheit (°F). The Celsius scale is most commonly used by climatologists and in almost all other natural sciences, and Celsius is used throughout this text unless otherwise stated. The origin of the Celsius scale and how it differs from others is explained in box 1.1.

The Earth's north-south variation in temperature is very strong, ranging through more than 90°C from the coldest region (Antarctica in winter) to the hottest (North Africa in summer). This range is more than twice as large as the seasonal temperature variation experienced anywhere on Earth. The global average near-surface air temperature is close to 15°C (60°F), a temperature that typifies the average conditions of middle latitude temperate regions of North America and much of Europe in the spring and fall.

BOX 1.1 TEMPERATURE MEASUREMENT SCALES

Fahrenheit. First proposed in 1724 by Dutch-German physicist Daniel Fahrenheit, the zero point represents the temperature of a specific brine solution, and 100 degrees is approximately human body temperature (although, in reality, the average is 98.6). Fahrenheit remains the official scale of the United States, Belize, and parts of Canada.

Freezing point of water (at 1 atmosphere): 32°F
Boiling point of water (at 1 atmosphere): 212°F

$$°F = °C \times (9/5) + 32$$
$$°F = K \times (9/5) - 459.67$$

Celsius. The Celsius scale is named after Swedish astronomer Anders Celsius, who developed a similar scale two years before his death in 1744. His scale was "reversed," and numbers decreased with heat, but the simplest definition of the scale remains the same: 0°C is the freezing point of water (at 1 atmosphere), and 100°C is the boiling point (at 1 atmosphere). Celsius is the most common temperature scale used across the globe.

Freezing point of water (at 1 atmosphere): 0°C
Boiling point of water (at 1 atmosphere): 100°C

$$°C = (°F - 32) / (9/5)$$
$$°C = K - 273.15$$

Kelvin. In 1848, William Thompson (later Lord Kelvin) developed a thermodynamic scale independent of measuring material. One Kelvin (unlike Fahrenheit and Celsius, measurement in the Kelvin scale is not referred to in degrees) has the same magnitude as one degree Celsius; however, absolute zero, the hypothetical but unattainable temperature at which matter exhibits zero entropy, is defined as being precisely 0 K (−273.15°C). The Kelvin scale is the primary unit of measurement in the physical sciences.

Freezing point of water (at 1 atmosphere): 273.15K
Boiling point of water (at 1 atmosphere): 373.15K

$$K = (°F + 459.67) / (9/5)$$
$$K = °C + 273.15$$

Other scales. In addition to Fahrenheit, Celsius, and Kelvin, a number of other temperature measurement scales exist. Rankine, proposed by a physicist of the same name in 1859, defines zero as absolute zero and a degree as equal to one degree Fahrenheit (which Kelvin defines as Celsius). Others include Réaumur, Rømer, and Delisle.

Earth revolves around the Sun in a slightly elliptical orbit, with the Sun at one focus. In figure 1.6, the *ellipticity* is greatly exaggerated. Earth rotates on its axis tilted about 23.5° away from vertical,[1] and it maintains that tilt throughout the complete passage around the Sun. Thus, in the northern summer, the Sun's energy strikes Earth most directly in the northern hemisphere, making that hemisphere at that position in the orbit warmer, as seen in figure 1.6 with Earth on the left. The northern winter/ southern summer occurs when Earth is on the right of the Sun. The arrow pointing from the Sun maintains its orientation, and to an observer on Earth, it would appear to move annually between the tropics of Capricorn in the south and Cancer in the north. This will become important in understanding the

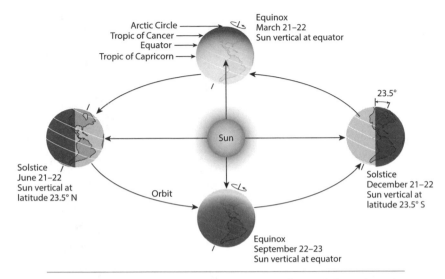

FIGURE 1.6 The tilt of the Earth's axis is maintained at 23.5° throughout the orbit around the Sun, causing the Sun to strike the Earth directly in different hemispheres depending on the position in the orbit.

apparent motion of seasonal rainfall in Africa and the Indian monsoon (see chapter 3). The surface traced out by Earth's orbit is known as the *plane of the ecliptic*.

WHY 15°C (60°F)?

The temperature experienced at Earth's surface is almost entirely the result of heating by radiation from the Sun. Although the deep interior of Earth is extremely hot (the temperature at the outer surface of Earth's core is approximately the same as that of the Sun's outer surface), the core's heat, as perceived at Earth's surface, although measurable, is insignificant in comparison.

In that sense, Earth can be considered thermally passive, whereas the Sun is thermally active.

Like any thermally passive object heated by an external source, Earth absorbs energy radiated from the Sun and, in equilibrium, reradiates the same amount back into space. This is known as *radiative balance.* The temperature at Earth's surface is achieved through the balance of absorption and reradiation of the Sun's energy. Four main factors contribute to establishing Earth's energy balance and give Earth its average temperature of 15°C (60°F):

1. *Solar energy output.* Were the Sun to increase in intensity with all others factors remaining the same, Earth would receive and absorb more energy, reach a higher temperature, and radiate more energy back into space.

2. *Geometric effects.* Incoming energy from the Sun is diminished in intensity in its passage across space to Earth, and having reached Earth, the energy is spread over the planet's surface.

3. *Earth's albedo.* Only energy absorbed by Earth causes it to gain heat. A fraction of the Sun's energy is reflected back into space, is not absorbed, and does not contribute to heating Earth's surface.

4. *Greenhouse effect.* Like the solid Earth, the *atmosphere* can absorb and reradiate heat energy because it contains *greenhouse gases* (GHG). If the fraction of GHGs in the atmosphere changes, the temperature at Earth's surface also changes to reestablish radiative balance.

The third and fourth of these factors can change and have been changed by human activity.

Factor 1: Solar Energy Output

The Sun's temperature is deduced from the nature of its emitted radiation, specifically its *electromagnetic spectrum*, using principles derived from atomic physics that were developed well over a century ago, mainly in Europe, during the birth of quantum mechanics (see box 1.2). Electromagnetic radiation from the Sun propagates as a continuous wave of energy that is created by the vibration of charged particles in the Sun. It is termed *electromagnetic* because the vibration of electrons creates an electric field that then generates a magnetic field; this statement could also be made in reverse. This is not unique to the Sun; all hot objects produce radiation in the same way. An incandescent or halogen lamp creates light energy because the active element is very hot. These two types of lamps have quite different spectra because halogen lamps emit at a higher temperature. LED (light-emitting diode) lamps operate on a different principle.

All waves, including electromagnetic waves, can be described by two properties: wavelength and amplitude. The wave shown in figure 1.7 is propagating electromagnetic energy from left to right (although it could equally have been propagating right to left). A sound wave that transmits acoustic energy can be characterized in the same way. It is important to understand that energy is being propagated—not an object, such as a surfer moving on an ocean wave. The amplitude of the wave is defined as the height above a middle line (not the full height from peak to trough). Higher amplitude waves carry more energy—high amplitude sound waves are louder. Radio waves are also a form of electromagnetic radiation; turning up the volume on a radio increases the amplitude of radio waves and hence increases their loudness. The sound is audible to us because the speakers convert the electromagnetic energy to sound waves.

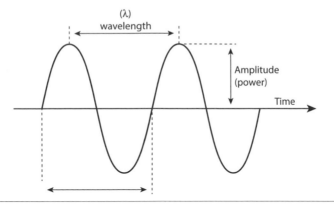

FIGURE 1.7 An electromagnetic wave propagating from left to right.

Wavelength, designated λ, is the distance between two peaks (or two troughs). Frequency is directly related to wavelength as $1/\lambda$, and it measures the number of peaks that pass by a fixed point in a defined period of time, usually 1 second. Higher frequency waves have more peaks passing by per second than do low frequency waves. Frequency is described in units of *Hertz* (Hz).[2] The inverse relationship of frequency and wavelength means that long wavelengths are associated with low frequency, and short wavelengths are associated with high frequency. For sound waves, low frequency waves produce bass notes, and high frequency waves produce shriller notes, such as a tuba compared to a piccolo.[3]

All electromagnetic waves, regardless of frequency or amplitude, travel at the speed of light in a vacuum, usually denoted as "c" (unfortunately, the same symbol is used for degrees Celsius, although for temperature the "C" is usually capitalized). On your radio, changing the station means changing the wavelength of the signal you wish to hear. Radio waves are very low frequency and have long wavelengths. While traveling through the vacuum of space, the frequency characteristics of waves remain unchanged.

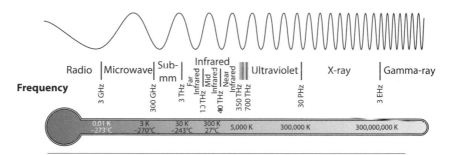

FIGURE 1.8 The relationship between temperature and frequency characteristics of electromagnetic waves, with some familiar waves noted.

Figure 1.8 illustrates the frequency/wavelength range of some familiar electromagnetic waves. For radio waves, about a hundred thousand peaks arrive at your radio each second, and for X-rays, 10,000,000,000,000,000,000 (or 10^{19}) peaks penetrate your fleshy matter and bounce off your bones each second. The visible part of the spectrum is shown expanded and has colors ranging from red through violet. The colors correspond to wavelengths from 400 to 700 nanometers (nm), a billionth of a meter. A glass prism can separate light into its constituent colors, and a rainbow forms under special conditions that create a type of atmospheric prism.

On either side of the visible band are infrared and ultraviolet waves, neither of which can be perceived by human eyes. Infrared radiation is often described as "heat," but that is not accurate. Many animals are able to see a little on either side of the visible spectrum; bees are able to see quite well into the ultraviolet. Night vision goggles used by the military sense radiation in the infrared range and respond to the heat of an object, so they can sense the presence of a person (or any warm-blooded animal) in the pitch dark.

Wien's law and the Stefan-Boltzmann relationship are essential formulas that relate the properties of electromagnetic radiation to the temperature of the emitting body and to the intensity of radiation produced by the body. These physical laws were developed in the early years of atomic physics and have stood the test of time. Key scientists in their development were Wilhelm Wiens (1864–1928), Max Planck (1858–1947), Ludwig Boltzmann (1844–1906), and Joseph Stefan (1835–1893). Their findings were known to science by the end of the nineteenth century or the early twentieth century.

The key insight is the *Planck function*, which is the derived relationship between intensity of radiation from a *blackbody* and the wavelength of the radiation (see box 1.2).

BOX 1.2 THE PLANCK FUNCTION

The Planck function describes the shape of the spectrum of a blackbody emitting radiation at any given temperature. Hotter bodies have taller peaks than those from cooler bodies.

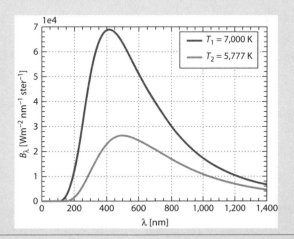

BOX FIGURE 1.2.1 The Planck function.

Source: Modified from Guðmundur Kári Stefánsson, "Plotting Planck's Law," January 18, 2014, http://gummiks.github.io/2014/01/18/planck/.

The peak is also moved left to shorter wavelengths/higher frequency. If the spectrum of an emitting blackbody can be measured, the temperature of that body can be determined.

"Blackbody radiation" or "cavity radiation" refers to an object or system that absorbs all radiation incident upon it and reradiates energy, which is characteristic only of this radiating system and is not dependent upon the type of radiation incident upon it (http://hyperphysics.phy-astr.gsu.edu/hbase /mod6.html).

Wien's law tells us that the temperature T of a body can be found from the peak of the Planck function:

$$\lambda_{max} = \frac{a}{T}$$

where a is a constant. It is obtained by differentiating the Planck formula and setting the derivative to zero. The *Stefan-Boltzmann* relationship gives the intensity I of the radiation that is related to the fourth power of the temperature and is given by the integral of the Planck formula over all wavelengths.

$$I = \sigma T^4$$

where σ is also a (different) constant.

With the formulae in box 1.2, the temperature and energy flux from the Sun can be calculated from its electromagnetic spectrum, assuming it radiates in the manner near to that of a blackbody, which is a very close approximation to the actual radiation properties of the Sun.

Factor 2: Geometric Effects

If the energy output of the Sun remains fixed (it varies a little today but has varied as much as 25 percent in earlier times), the energy received at the top of the atmosphere depends completely on the distance separating Earth from the Sun. A vacuum is present in the space between these two bodies, so no energy is absorbed in the traverse across space. This relationship is not linear, however; if Earth were 50 percent farther from the Sun, it would receive much less than 50 percent of the Sun's energy. The reduction in energy received at increasing distance is due to a phenomenon called *spherical spreading*, or spherical divergence (sometimes called geometric spreading), which is the same way sound energy diminishes in strength with distance from a sound source.

Imagine an instantaneous pulse of energy emitted by the Sun. It would form an infinitely thin spherical sheet of energy moving out into space, expanding equally in all directions (figure 1.9).

The total area of the expanding shell of energy is $4\pi r^2$ for energy that has traveled a distance r from the Sun. Because of the r^2 factor, doubling the distance to $2r$ produces a sphere with

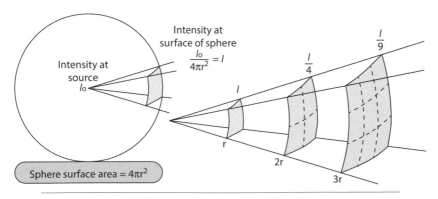

FIGURE 1.9 Radiation from the Sun showing the expansion of a unit area, A.

4 times the initial surface area. Triple the distance to $3r$ and the surface area is 9 times larger. In figure 1.9, this is illustrated with an element of the sphere with an area A at distance r. When the distance has doubled, the area becomes $4A$, when tripled, it is $9A$. Think of inflating a balloon a small amount, drawing a square on it, and then continue inflating. The square will get larger and larger. The formula tells us how much larger it gets for a given change in radius.

The energy density—the amount of energy in a square kilometer—diminishes according to this simple formula, the general form of which is known as an inverse square law:

$$E_r \propto \frac{E_0}{r^2}.$$

This equation says that energy at a distance r from the Sun, E_r, is proportional to the energy of the Sun, E_o, derived from the Planck function (see box 1.2), divided and therefore reduced by the square of the distance r^2.

Using an average distance of 150 million kilometers (km), the inverse square law of spherical divergence tells us that the Sun's energy density (measured in watts per km^2) has been reduced by a factor of more than a billion by the time it reaches Earth; i.e., only a billionth of the Sun's radiated energy is available to warm Earth. The remainder spreads out evenly into space in all directions. It is important to note that the change in energy in the passage through space involves only the amplitude of the propagating energy, and not the frequency.

A related form of spreading the Sun's energy accounts for most of the change in temperature from the equator to the poles. In figure 1.10, the lines represent rays of energy from the Sun. Because the Sun is so distant, the rays can be considered to be essentially parallel. Rays are not real, of course; they are commonly used geometric constructions that help illustrate certain

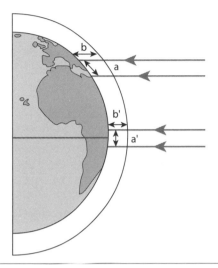

FIGURE 1.10 The Sun's rays are overhead at the equator, but always impinge at an angle at other locations. This causes the Sun's energy to spread over greater areas at higher latitudes.

phenomena, particularly those associated with the propagation of energy over distance.[4]

In the equatorial region, these rays strike Earth at 90° to the surface in spring and fall (see figure 1.6). To an observer at or near the equator, the midday Sun appears directly overhead, and the heat is intense. Toward the poles, the Sun's rays strike Earth at an increasingly smaller angle. Something like this is experienced at sunset. Shadows get longer as the Sun lowers in the sky and the air temperature becomes cooler. At very high latitudes, the Sun is always in a position similar to the setting Sun in lower latitudes.

Consider energy carried to Earth between any two of the Sun's rays. Because of the curvature of Earth, energy is spread over a much larger area, a, near the pole than the area, a', at the equator. The analogy of a flashlight shining on a flat surface is often used to describe this idea (figure 1.11).

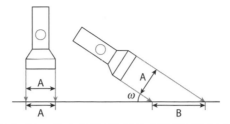

FIGURE 1.11 A simple way to consider the effect of decreasing temperature is to use a flashlight (as shown on the left) in which the radiated area (*A*) illuminates the same area on the surface. When the flashlight is tilted to an angle *ω* (as shown on the right), the same amount of energy (*A*) from the flashlight is spread over a greater area (*B*).

The factor by which the area increases is $1/\text{cosine}(\omega)$, where ω is the angle of latitude (or the tilt of the flashlight). So one square meter of Iceland at latitude 60° N receives around half as much of the Sun's energy as does one square meter of Lake Victoria in Kenya, at the equator, and it is this that accounts for the difference in average temperature between Nairobi and Reykjavik (not that Reykjavik is farther from the Sun than Nairobi). Another important consequence of the cosine factor is that conditions vary much more slowly in the tropics than in other regions of Earth. If you travel 130 miles in any direction from Nairobi, average conditions will be almost exactly the same as long as you stay at the same altitude. But average temperatures between New York City and Albany, New York (130 miles apart), are distinctly different—winters, in particular, are much colder in Albany.

In addition, the thickness of the atmosphere appears to an observer on Earth, looking directly toward the Sun, to increase from b' to b (see figure 1.10). The atmosphere itself has not increased in thickness, although it appears to do so because the observer is looking straight at the Sun lower in the sky. The redness of sunrise and sunset is caused by scattering in the atmosphere, which is stronger at the blue end of the spectrum, so only red wavelengths remain after a long passage through the atmosphere. These two geometric effects—spherical spreading of energy in space and the spherical geometry of Earth—explain how much energy from the Sun arrives on Earth and account for the strong temperature gradient from equator to pole. This pattern is described as incoming solar radiation or *insolation*, a pseudo-acronym not to be confused with "insulation," and is the energy available to warm Earth.

Factor 3: Earth's Albedo

Earth does not absorb 100 percent of the incoming solar energy; a significant fraction is reflected back into space in the same way that light is reflected from the shiny surface of a mirror. In climate studies, the reflectivity of Earth is referred to as *albedo* (the Latin root means "whiteness"). In general, surfaces that appear white reflect more than darker surfaces. The angle at which the light strikes a surface is important as well; light striking at a low angle (when the Sun is low to the horizon) is reflected more completely than light striking from nearly overhead. Figure 1.12 shows albedo measurements derived from satellite observations.

Clouds contribute significantly to Earth's albedo. The common experience of being at the beach and feeling warm in the direct rays of the Sun, then feeling distinctly cooler when the "Sun goes behind a cloud," is clear evidence of the shielding properties of clouds. The clouds are reflecting energy back into space; clouds absorb some energy, but the primary effect is reflection for this type of cloud. The narrow band of high albedo across the equatorial Pacific and Atlantic is a cloud band associated with the *Intertropical Convergence Zone* (ITCZ; whose origin is described in chapter 2).

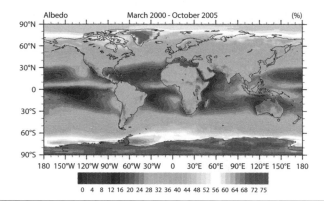

FIGURE 1.12 NASA CERES (Clouds and Earth's Radiation Energy system) satellite–derived measurements of Earth's radiation and albedo. Lighter areas are more reflective. Scale is percentage of sunlight reflected.

Source: "Mean Annual of Terrestrial Radiation [TOA] and Albedo," compiled and modified by Sandro Lubis, *Tropical Dynamics and Equatorial Waves* (blog), April 29, 2012, https://sandrolubis.wordpress.com/2012/04/29/mean-annual-of-global-outgoing -terrestrial-radiation-from-toa-clear-sky/.

Ice-covered areas are the most reflective Earth surfaces, achieving values near complete reflection. This is one reason ice-covered regions tend to remain cold—they benefit very little from the Sun's warmth. Cloudless oceans are strongly absorbent. Deserts are some of the most reflective land surfaces, which explains why they are often very cold at night; deserts absorb very little incoming radiation, so little is stored to provide warmth by reradiating at night. The air in desert regions is also very dry, and this enhances the nighttime long wavelength cooling of desert surfaces. Figure 1.13 shows the albedo factors in percentages— snow, for instance, can reflect almost 85 percent of the energy that impinges on its surface. It is important going forward to remember that different clouds have different albedo properties and that natural forests and croplands also differ in albedo properties.

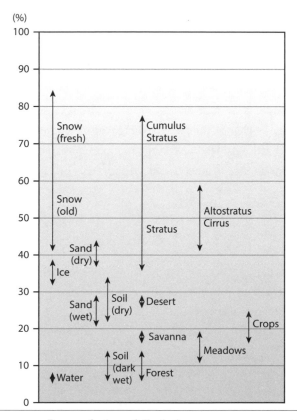

FIGURE 1.13 Range of values of albedo for some common materials
and surface conditions.

Source: Courtesy of Hannes Grobe/Wikimedia Commons.

The average total albedo over Earth's entire surface and atmosphere is approximately 0.3, so about 30 percent of incoming
energy from the Sun reflects back into space and plays no part
in warming Earth. The pattern of albedo varies with the seasons,
and natural events such as strong dust storms can cause changes
in total albedo as well.

Almost all human activity changes Earth's albedo to some extent, increasing or decreasing it, but not by a significant amount with respect to the global average. Land use change, such as deforestation and the corresponding replacement by cultivated lands, and urban growth have important local effects, but neither alter the global average albedo significantly, at least not enough to noticeably change the global average temperature. Any human activity that might lead to change in cloud cover could induce a significant change in albedo, and I discuss this in a later chapter.

Factor 4: The Greenhouse Effect

The interdependence of temperature, wavelength, and energy derived from atomic physics allows a calculation to be made of the temperature Earth would attain if it were simply heated by the Sun—that is, with no consideration of the atmosphere other than its cloud albedo properties. The basic calculation is shown in box 1.3, and the steps are as follows:

- The temperature of the Sun is obtained using Wien's law relating the wavelength of electromagnetic energy (measured with a spectrometer) to the temperature of an energy emitting body.
- Given this temperature, the Stefan-Boltzmann relation is used to calculate the total energy originating from the Sun.
- The energy at the top of the atmosphere is obtained by diminishing the energy from the Sun using a spherical divergence correction. That figure comes out to be 1,367 watts/square meter.
- Some simple spherical geometry is then needed to calculate how energy arriving from the Sun spreads over the spherical Earth. This leads to a figure of around 350 watts/square meter as the average incoming energy over the entire Earth's atmosphere.

- An average albedo of 0.3, meaning 70 percent of the energy arriving from the Sun, is then used in calculating the warming of Earth.
- Assume that the Earth-Sun system is in equilibrium and that Earth has been fully warmed by the Sun's heat and emits back into space an amount of energy equal to the amount it receives. Then the Stefan-Boltzmann relation is used again, this time in reverse, to calculate Earth's emission temperature.

Employing these six steps, the temperature of the planet is estimated to be about 16°C *below* freezing, much colder than Earth's true average temperature (about 15°C above zero), and more like the temperature in polar regions. The difference is attributed to the warming effect of the atmosphere (discussed later), but it is worth asking whether other factors could account for the difference.

BOX 1.3 CALCULATING EMISSION TEMPERATURE AND THE GREENHOUSE EFFECT

With respect to the figure for symbol definition:

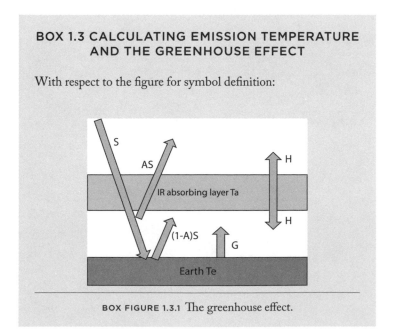

BOX FIGURE 1.3.1 The greenhouse effect.

Solar radiation S absorbed = planetary radiation emitted at emissions temperature Ta.

A is the albedo factor, R is the radius of Earth. To balance incoming radiation that arrives at Earth with that emitted we have:

$$S (\text{1-}A) \, \pi R^2 = \sigma(Ta)^4 \, .4\pi R^2 \text{ i.e. } I_{in} = I_{out}$$

where the terms in π account for the spreading of a disc of incoming energy over a spherical Earth. Then

$$(Ta)^4 = S (\text{1-}A)/(4\sigma)$$

using $A = 0.3$; $S = 1370$ W/m^2; $\sigma = 5.67 \; 10^{-8}$ W/m^2/K^4.

This gives Ta of $-16°$C, far too cold.

Including the IR absorbing layer H is the energy emitted and reradiated toward Earth, sometimes referred to as back radiation. So,

$$2 \, \sigma \, Ta^4 = \sigma \, Te^4 \quad (H + H = G) \quad G \text{ is the radiation from Earth}$$

where Te is Earth's emissions temperature and Ta that of the atmosphere.

Short wave radiation absorbed + layer radiation emitted = Earth radiation emitted

$$S (\text{1-}A)/4 + \sigma \, Ta^4 = \sigma \, Te^4$$
$$Te = 2^{(1/4)} \, Ta \text{ or } 1.19 \; Ta \; (S(\text{1} - A) + H = G).$$

Substituting previous results

$$Ta^4 = S (\text{1-}A)/(4\sigma)$$

then using $A = 0.3$; $S = 1370$ W/m^2 as before and Ta at $-16°$C, Te becomes about $30°$C.

1. Top of the Atmosphere Balance

   Solar radiation absorbed = planetary radiation emitted

   $$S\,(1\text{-}A)\,\pi\,R^2 = \sigma\,Ta^4\,4\pi\,R^2$$
   $$Ta^4 = S\,(1\text{-}A)/(4\sigma)\;(S(1\text{-}A) = H\,).$$

2. Atmosphere Layer

   $$2\,\sigma\,Ta^4 = \sigma\,Te^4 \quad (H + H = G).$$

Wien's law and the Stefan-Boltzmann relation have been experimentally verified numerous times. The distance from the Sun is also accurately known. It is possible that the average albedo is different from 0.3, but to account for the calculated temperature difference, albedo would need to be significantly higher (Earth would need to be much more reflective) so less heat from the Sun warmed Earth. In fact, if Earth really had no atmosphere, it would have no clouds, and that would lower the albedo. There is some uncertainty in the average value of albedo, but it is not sufficient to account for the difference in calculated temperature. In fact, if albedo were ignored and all incoming energy warmed Earth, the calculated temperature would only rise to about $-5°C$ ($20°F$). Finally, we can be confident that Earth does not behave exactly like a true blackbody to which the Planck function applies. It is often referred to as a gray body. However, it is not so different that it could account for the very cold temperature estimate.

What is missing in the previous calculation are the thermal properties of the atmosphere. The first speculations about the effect of the atmosphere in modulating Earth's temperature most likely came from French mathematician Joseph Fourier (1768–1830), who in the 1820s correctly deduced that the atmosphere must play an important role in determining Earth's surface temperature. This was well before the details of atomic physics had been worked out. The very low temperature that was derived from this first calculation can be thought of as Earth's blackbody temperature. In that sense, it is not incorrect. Following Fourier in the late nineteenth century, Svante Arrhenius (1859–1927), a Swedish physicist and physical chemist (the first Swedish Nobel Laureate, 1903), made a calculation of the effect on surface temperature of CO_2 concentrations and deduced that a doubling of concentration would raise surface temperature by 5–6°C, remarkably close to modern estimates.

The way to think of this is to recognize that Earth, having been warmed by the Sun, emits electromagnetic energy back into space, albeit at a much lower temperature, with a much longer wavelength (lower frequency), and with much less intensity than the Sun. This radiation is emitted back into space (in all directions), where it encounters the atmosphere. The atmosphere absorbs some of that energy and is itself warmed by Earth's radiation in the same way the Sun's heat warmed Earth. In the first calculation, all of Earth's heat was allowed to escape into space—clouds had no effect on the outgoing radiation. The atmosphere acts something like a warm covering, absorbing radiation emitted by Earth and radiating some of that energy downward (as well as out into space).

This effect is generally called the *greenhouse effect*, which is not a particularly good analogy because a true greenhouse literally traps heat inside (the glass roof of a greenhouse does not become heated and does not radiate energy back into the greenhouse). It is also sometimes said that the atmosphere "traps heat." This is only partly correct as well because the warming effect relies on the atmosphere absorbing and reradiating heat back to Earth, letting go what it trapped. One way the greenhouse effect is sometimes described that is quite wrong is to say heat energy from Earth is reflected back by the atmosphere. If that were true, the use of the greenhouse analogy would be appropriate. In fact, very little reflection of long wavelength radiation occurs from the atmosphere.

To complete the calculation of surface temperature, a seventh step is added to the previous sequence: Add heat that comes back to the surface from long wavelength radiation that is absorbed and reradiated by the atmosphere toward Earth—sometimes referred to as back radiation. That is shown at the bottom of box 1.3. The calculated temperature is now a lot closer to what is experienced, although it is higher than the true average because the only form of heat transfer considered in the calculations is by radiation. In fact, heat also moves in two additional forms— conduction and convection. Adding these two components leads to the correct temperature, but it is beyond the scope of this book to discuss the details of those effects.

THE MECHANISM OF GREENHOUSE WARMING

Reviewing the logic described here, something might seem not quite right. Three bodies are involved in emitting and receiving electromagnetic radiation: the Sun, solid Earth, and the atmosphere. In simple terms, the Sun warms Earth, and Earth warms the atmosphere. But why doesn't the Sun also warm the atmosphere from above? The atmosphere is composed of many gases, with oxygen and nitrogen comprising by far the greatest fraction, neither of which is a greenhouse gas (table 1.1).

Gases referred to as greenhouse gases (GHGs) are present in trace amounts. The present concentration of carbon dioxide is

TABLE 1.1 Major chemical constituents of Earth's atmosphere

Atmospheric gases	Chemical symbol	Percent by volume in the atmosphere	Concentration in parts per million (ppm)	Average residence time	Variability over time and spatial scales
Non–greenhouse gases					
Nitrogen	N_2	78.1	780,840	1.6×10^7 years	Constant
Oxygen	O_2	20.9	209,460	$3 \times 10^3–10^4$ years	Constant
Argon	Ar	0.9	9,340	N/A	Constant
Greenhouse gases					
Carbon dioxide	CO_2	0.0397	397	100–300 years	Variable
Methane	CH_4	0.000179	1.79	10–12 years	Variable
Nitrous oxide	N_2O	0.0000325	0.3	121 years	Variable
Ozone	O_3	0.00004	0.01–0.5	150 hours— days	Highly variable
Water vapor	H_2O	Variable 0.001 to 0.5, strongly varies locally	10–50,000	8–10 days	Highly variable
Chloro-fluorocarbons (CFCs)	Contain C and H	0.0000002	0.0002	9 years to 3,200 years depending on type of CFC molecule	Highly variable

Source: Intergovernmental Panel on Climate Change, *Climate Change 2007: Synthesis Report*, contribution of Working Groups I, II, and III to the *Fourth Assessment Report of the Intergovernmental Panel on Climate Change*, ed. Core Writing Team, Rajendra K. Pachauri, and Andy Reisinger (Geneva: Intergovernmental Panel on Climate Change, 2007).

about 0.040 percent or 400 *parts per million* (ppm), and methane is measured in *parts per billion* (ppb). Water vapor is also a powerful greenhouse gas; it is more abundant than others but still constitutes less than 0.5 percent of the atmosphere. Those greenhouse gases referred to as anthropogenic GHGs are a subset of trace gases that are due to or influenced by human activity. Water vapor, slightly influenced by human activity, is not usually considered an anthropogenic GHG. In 1869, John Tyndall confirmed the warming effects of some of these gases (water vapor and carbon dioxide) through a set of laboratory experiments. How does this warming effect work?

Again, the answer lies in basic atomic physics. When electromagnetic energy interacts with gases at relatively low temperature, gas molecules absorb energy by vibrating more vigorously. This is the inverse effect of what was previously described, in which the vibration of charged particles produced electromagnetic radiation. Because each gas molecule has a different size, shape, and molecular structure, and the bonds that hold their atoms together have different strengths, the amount of energy absorbed by each molecule differs with the wavelength of radiation impinging on them. Many molecules have several modes of vibration, each one responding to a different wavelength of the impinging radiation. Figure 1.14 is a stylized depiction of carbon dioxide showing its four vibration modes.

The carbon atom is depicted as larger than the oxygen atoms, and they are connected by atomic bonds. Arrows indicate the relative direction of vibrating motion of the atoms. Each mode is excited by electromagnetic energy of a different wavelength. That means that one carbon dioxide molecule has four ways to absorb energy, so, in effect, it acts like it is four different molecules.

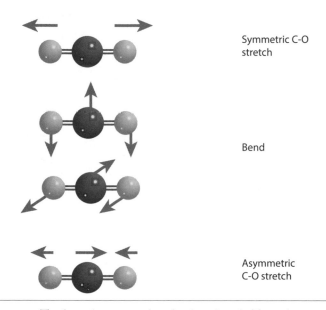

Symmetric C-O
stretch

Bend

Asymmetric
C-O stretch

FIGURE 1.14 The four vibration modes of carbon dioxide. The carbon atom
is represented by the larger, darker atom.

Generally, molecules with three or more atoms have several vibration modes and are effective at absorbing electromagnetic radiation at several different wavelengths. All of the GHGs have multiple modes of vibration. Those gases with molecules composed of two atoms—such as nitrogen and oxygen (N_2 and O_2), which make up most of the atmosphere—are tightly bonded. They do not vibrate vigorously, so they do not absorb as effectively. They have only one vibration mode and are largely insensitive to long wavelength radiation. Nitrogen, in particular, has one of the strongest bonds of all molecules, and thus, despite the large concentration in the atmosphere, plays no part in the greenhouse effect. Conversely, complicated multiatom long-chain molecules such as chlorofluorocarbons have numerous vibration modes and have very strong greenhouse effects.

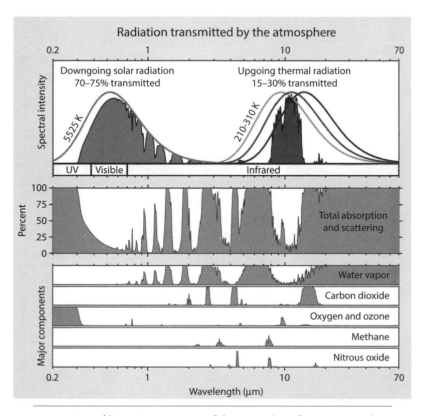

FIGURE 1.15 Absorption spectrum of the atmosphere for incoming solar (short wavelength) radiation and outgoing (long wavelength) terrestrial radiation.

Source: Courtesy of Robert A. Rohde / Wikimedia Commons.

The aggregate effect of molecular absorption is depicted as an *absorption spectrum* in figure 1.15. Each gas molecule has a unique absorption spectrum that is a consequence of its individual vibration modes, as described previously for carbon dioxide. The total effect of all gases in the atmosphere is added together and shown in the figure, with individual absorption bands of water vapor, oxygen, ozone, methane, and carbon dioxide beneath.

At the top, in solid lines, are emissions spectra of the Sun (at left) and Earth (at right) unmodified by interaction with the atmosphere. The Sun's energy peaks at a very short wavelength in the visible range, and Earth's radiation spectrum peaks at a long wavelength in the infrared range, because the Sun is very hot (5,525K or 9,500°F; box 1.1 explains the Kelvin temperature scale) and Earth is relatively very cool (spectra for a range from 210K to 310K [100°F]).

The jagged in-filled area at the left is the spectrum of solar radiation that reaches the surface of Earth after passing through the atmosphere. It is modified somewhat, mainly due to the absorbing properties of ozone (O_3) at short wavelengths and by water vapor at somewhat longer wavelengths, but most of the Sun's short wavelength energy penetrates to reach Earth's surface with little modification to its spectrum. The atmosphere, through its reflective properties, does influence the amplitude of the incoming radiation—here the discussion concerns the energy spectra. The effect of ozone on short wavelength absorption is extremely important because it provides protection from harmful ultraviolet radiation.

The jagged in-filled area at the right is the much-modified spectrum of Earth's long wave radiation after it has interacted with the atmosphere. Carbon dioxide and water vapor are responsible for most absorption, with methane (CH_4) and nitrous oxide (N_2O) contributing smaller effects. The energy that does not make it through the atmosphere is absorbed and reradiated, some outward into space and some back toward Earth—this is the greenhouse effect.

Carbon dioxide has no absorption band at short wavelengths, so it has no effect on short wavelength incoming solar radiation. In contrast, it is very effective at absorbing outgoing long wavelength infrared radiation emitted by Earth because it has strong absorption bands at long wavelengths. Fundamentally, the greenhouse effect arises because of the way in which electromagnetic radiation of different wavelengths interacts with a large suite of gases in the atmosphere.

As an everyday analogy, think of the effect of a microwave oven on different foods and containers. Microwaves are a form of electromagnetic radiation (see figure 1.8). Place a slice of apple crumble in the oven in a glass container, or a slice of last night's pizza on a paper plate, and run the oven for a few minutes. The crumble will get warm, but the glass will not. Why? The answer is that the molecular structure of apple crumble is different from that of glass, and it reacts differently to microwave radiation. Pizza molecules are different from paper plate molecules, and pizza crust is not the same as pizza sauce or pepperoni. The glass is like oxygen in the atmosphere, and the crumble is like carbon dioxide. The reason metal objects placed in a microwave oven spark vigorously is because metals react very strongly to microwave radiation.

The takeaway message from this discussion is that the phenomenon known as the greenhouse effect can be explained by quantum mechanics principles established more than a century ago, and they are indisputable. It is this that explains the average global temperature of Earth. In the absence of a greenhouse effect, Earth would be much colder, on average, and only a small fraction would be habitable by humans. The greenhouse effect is entirely an atmospheric phenomenon dependent on the presence of certain trace constituents of the atmosphere known as greenhouse gases.

2

PRECIPITATION, WINDS, ATMOSPHERIC PRESSURE, AND THE ORIGIN OF CLIMATE ZONES

S OME places on Earth persistently receive more rainfall than others, and some are much windier than others. Water is essential for almost all life functions, and even the most technologically advanced agricultural regions in the world today rely on rainfall in the right amount at the right time for successful crop yields. In some regions, winds blow consistently from east to west; in others, just the opposite. These features cannot be explained by patterns of temperature alone, our first-order approximation (described in chapter 1). This chapter provides the next refinement to understanding the climate system, our second-order approximation, in which we account for patterns of wind and rainfall. These refinements can be explained using classical Newtonian mechanics.

PATTERNS OF WIND, PRECIPITATION, AND SEA LEVEL PRESSURE

Figures 2.1 and 2.2 show the pattern of surface winds in the northern summer and winter. Arrows indicate wind direction, and the shaded scale indicates wind strength, with darker areas representing stronger winds.

JJA

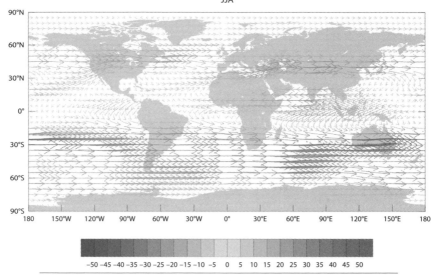

−50 −45 −40 −35 −30 −25 −20 −15 −10 −5 0 5 10 15 20 25 30 35 40 45 50

FIGURE 2.1 Global wind field in northern summer. ("JJA" refers to the months of June, July, August.)

Source: National Center for Environmental Prediction (NCEP).

DJF

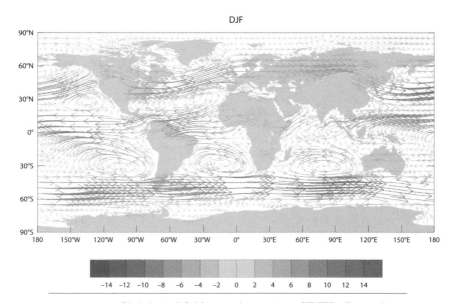

−14 −12 −10 −8 −6 −4 −2 0 2 4 6 8 10 12 14

FIGURE 2.2 Global wind field in northern winter. ("DJF" refers to the months of December, January, February.)

Source: National Center for Environmental Prediction (NCEP).

Similar to temperature, wind patterns are strongly zonal, with much stronger variations in a north-south direction than east to west. Bands of strong wind alternate direction rather than gradually changing from the equator to the poles. The direction of wind flow is predominantly east to west or west to east, and wind flow changes direction 180° from one zonal band to the next. For example, the strong winds in the central Pacific flow from east to west, whereas the strongest band of wind in the southern ocean flows in the opposite direction, west to east. The central Pacific winds, known as the northeast trade winds, were critically important to ocean travel under sail in the early days of global commerce. The areas of very low wind strength—about 30° in both hemispheres—are known as the *horse latitudes*. Sailing ships were often becalmed in these areas and would lighten their ships to catch whatever faint winds there were by throwing heavy items overboard, including horses as a last resort.[1] The *doldrums* is the area of slack winds around the equator. The strength and intensity of winds change with the seasons, but wind direction remains largely unchanged (an exception to this is in El Niño years, discussed in chapter 3).

The pattern of sea level pressure (SLP) is also mostly zonal in overall form, but some features have a distinct east-west variation. For example, the low pressure system in the western equatorial Pacific and high pressure near the west coast of North America and South America have a distinct east-west variation. Figure 2.3 shows SLP averages over one year.

The last factor to be considered at this point is the pattern of precipitation shown in figure 2.4, also as an annual average.

The most striking feature of this map is the narrow band of high-intensity rainfall across the equatorial Pacific in the Indonesian region and through the Indian Ocean. This band was seen in chapter 1 as the strong albedo effect of clouds in the same

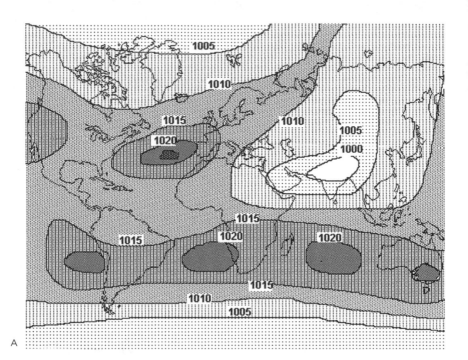

A

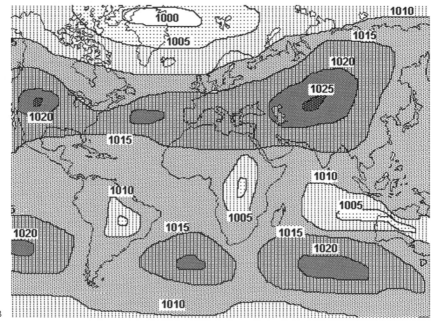

B

FIGURE 2.3 Global pattern of sea level pressure expressed as an annual
mean in (*A*) January and (*B*) July.

Source: Japan Meterological Agency, "JRA-55 Atlas (Japanese 55-Year Reanalysis Project),"
http://ds.data.jma.go.jp/gmd/jra/atlas/en/surface_basic.html.

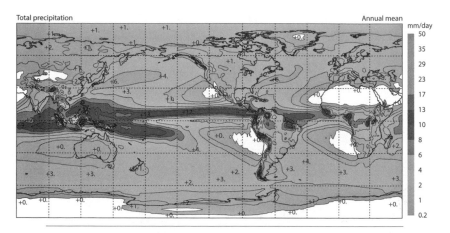

FIGURE 2.4 Global annual rainfall.

Source: K. E. Trenberth, John Fasullo, and Lesley Smith, "Trends and Variability in Column-Integrated Atmospheric Water Vapor," *Climate Dynamics* 24, no. 7–8 (2005): 741–758, https://doi.org/10.1007/s00382-005-0017-4.

region (see figure 1.12). Wind direction and strength, rainfall patterns, pressure, and temperature fields are all interconnected. In fact, these characteristics of the atmosphere, taken together, are what is usually referred to as Earth's climate.

ORIGIN OF THE OBSERVED PATTERNS

Earth's climate is influenced by two additional factors: air pressure gradients and rotational effects. First, let's examine air pressure gradients. SLP can be thought of as the weight of the air mass above the surface. If air is less dense, it weighs less, creating lower air pressure. For example, as the air in the equatorial region is heated, it expands and becomes less dense. The Antarctic continent, the coldest place on Earth, has a very dense atmosphere,

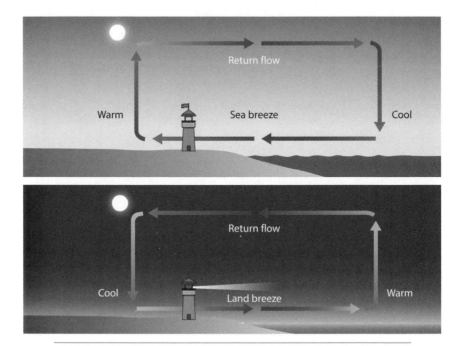

FIGURE 2.5 The development of a sea breeze due to differing pressure at sea and on land caused by differential heating.

resulting in high atmospheric pressures. The flow of air from high to low pressure regions works toward equalizing the pressure differences in an attempt to bring the system to equilibrium in a minimum energy state. This flow of air is what we call "wind," and a familiar example is the formation of a sea breeze (figure 2.5).

During a typical day, land absorbs heat (low albedo) much more quickly than the ocean does, causing air pressure to become higher over the sea. This creates a landward flow of relatively cool air that blows in off the ocean, which beachgoers appreciate. Aloft in the atmosphere, a return flow of air moves in the opposite direction to balance the air masses, with the whole system forming a circular loop (upper panel of figure 2.5).

This effect often reverses at night (lower panel of figure 2.5). The ocean cools more slowly because it retains more absorbed heat than the land (similar to the cooling of a desert at night). This may create a warm breeze flowing from the land toward the ocean, which is referred to as a land breeze.

The strong equator-to-pole temperature gradient—around 90°C—gives rise to a global pressure gradient that creates a driving force for global airflow. This happens in both hemispheres, so global airflow resulting from this gradient alone might look something like the illustration in figure 2.6. This diagram is

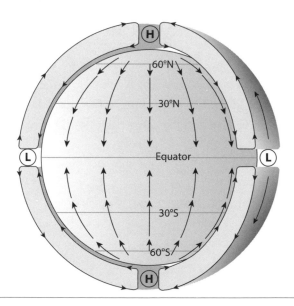

FIGURE 2.6 The global pattern of air circulation that would result from differences in temperature/pressure at the poles and the equator.

completely imaginary, however, because it assumes that Earth is not rotating and remains hot around the entire equatorial region. It is a way of asking what the pattern of airflow would look like were the equator-to-pole pressure gradient the only force involved in causing air to move. Only on a very slowly rotating Earth could such a pattern develop. The length of a day on Venus is equivalent to 243 Earth days, and Venus rotates around the Sun in 224.7 days, so a day on Venus is longer than its year. Venus does rotate very slowly, and it exhibits a single cell system of atmospheric motion much like that shown in figure 2.6.

Rotational effects are the second factor affecting Earth's climate. In 1735, George Hadley, an English lawyer and amateur meteorologist, first proposed the circulation pattern shown in figure 2.5 as a way to explain the trade winds. But it clearly cannot explain the observations in figures 2.2, 2.3, and 2.4, in which we observe strong alternating bands of airflow oriented east to west. North-south airflow is rare, with the exception of some local effects, so another force must act to move air east to west and west to east. That force is provided by Earth's rotation, which gives rise to the *Coriolis effect* (box 2.1). Named for Gaspard-Gustave Coriolis, who published a comprehensive paper on the subject in 1835, the effect had been

BOX 2.1 THE CORIOLIS EFFECT

The Coriolis effect refers to the deviation experienced by objects moving on a rotating body. In the figure, the object at A is attempting to go to B by moving due south. Viewed from directly over the North Pole, Earth rotates anticlockwise at a rate of around 15° longitude per hour.

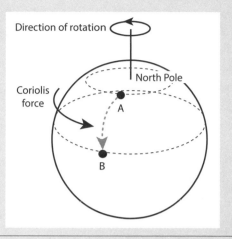

BOX FIGURE 2.1.1 The Coriolis effect.

The formula in scalar form for the Coriolis effect is $C = 2mV\Omega\sin\Phi$ where Φ is the angular latitude and Ω is Earth's angular velocity and m is the mass of the moving object moving at a velocity V. The latitude at the equator is zero, so the Coriolis effect is zero at the equator as well. The Coriolis effect alters the apparent direction of motion of the object, but not its speed of motion.

Imagine an object that starts out at the pole and moves relentlessly toward the equator. This would describe the intended motion of surface winds driven by an equator-to-pole temperature/pressure gradient. Because Earth rotates, the point being aimed toward on the equator moves to the left (relatively) as Earth rotates and the object continues on its straight line toward the equator. This happens because the wind is not well connected to Earth. If instead we considered an object being dragged along Earth's surface, this effect would not be felt. Earth essentially moves beneath the wind as the wind flows toward the equator. (In fact, there is some friction between surface winds and Earth's surface, so the wind's apparent path is not perfectly described

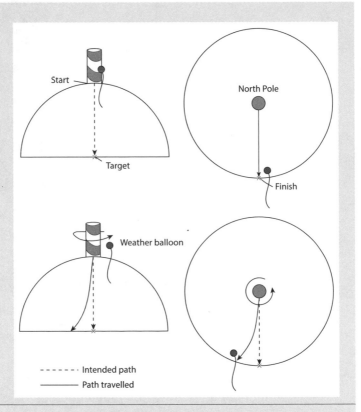

Start

Target

North Pole

Finish

Weather balloon

- - - - - Intended path

——— Path travelled

BOX FIGURE 2.1.2 The Coriolis effect on a moving object.

by the Coriolis effect.) Aircraft need to account for the Coriolis effect when making long-distance flight plans because they, too, are not strongly coupled to Earth's surface. Trying to hit an object at a great distance using a rifle also requires some consideration of the Coriolis effect. But this effect is weak when compared with gravitation or centrifugal forces and has no noticeable effect on most human activities. To be noticeable, Earth would need to be rotating at least 10 times faster than it does today.

For instance, think of an airplane flying due south from London, England, to Accra, Ghana, which is almost due south and about 5.5° N. If the pilot flies above the clouds so there is no ground reference, and navigates by keeping the compass pointing due south, Accra would have moved east with the rotating Earth before the plane arrived, and the plane might land in Sierra Leone instead (depending on the speed of the plane). If the flight were long, the plane might end in the Atlantic. Coriolis is a virtual or fictitious force, described more correctly as an "effect," because no force actually deflects the plane's path. On a rotating disc (a common way to explain the effect), it acts horizontally, but on a sphere it has horizontal and vertical *vector* components.

The Coriolis effect is not to be confused with the Eötvös effect, which is also due to Earth's rotation, and in this case is a true force experienced in the perceived change in the gravitational force on a body traveling east to west or west to east, i.e., with or opposed to the rotation of Earth and so either increasing or decreasing the angular velocity of the body.

The Coriolis effect causes deflection of moving air masses to the right in the northern hemisphere and to the left in the southern hemisphere.

recognized and described as early as the 1650s.[2] However, it was not developed in a mathematical form until the work of Coriolis. Hadley's model does not include the Coriolis effect, and he cannot be faulted for that because it was proposed one hundred years before Coriolis published his work.

THREE CIRCULATING ATMOSPHERIC CELLS IN EACH HEMISPHERE DETERMINE EARTH'S CLIMATE

Three large circulation cells are mirrored in each hemisphere. Figure 2.7 is a 3-D rendering of their pattern, including return flow in the upper atmosphere.

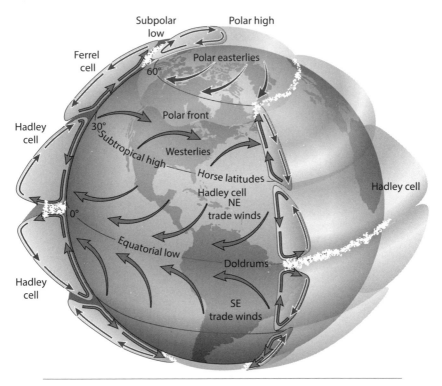

FIGURE 2.7 A depiction of the atmospheric circulation system, including return flow in the upper atmosphere.

Source: Based off of "NASA's Remote Sensing Tutorial: The Water Planet - Meteorological, Oceanographic and Hydrologic Applications of Remote Sensing."

Consider just the cells that meet at the equator, referred to as Hadley cells. These cells can be thought of as a very large version of a pressure-driven sea breeze circulation (see figure 2.5) that extends to about 30° N and about the same extent south. The northern and southern hemisphere Hadley cells meet at the equator, where air rises due to strong heating (recall figure 1.10). Because it is very hot at the equator, extensive evaporation and *evapotranspiration*[3] take place, so the rising air contains large amounts of water vapor. As this air rises, it cools and forms clouds of condensation, which create a band of rainfall across the equator. The condensation process itself also heats the atmosphere due to the release of *latent heat* when water vapor (a gas) changes phase to liquid water. The equatorial low pressures, the band of intense rain, and the band of high cloud albedo at the equator (see figure 1.12) are all associated with the upwelling of very warm, moist air at the junction of the two Hadley cells. The region where the two Hadley cells meet and air rises strongly is referred to as the Intertropical Convergence Zone (ITCZ). Figure 2.8 shows a satellite image of the cloud band associated with ITCZ storm systems.

The three-cell pattern of circulation comes about as a result of competition between the Coriolis effect and pressure-driven forces. Their relative size on our planet means that a body of air moving directly toward the pole from the equator in response to thermal stresses is deflected to the east-west after it has traveled through about 30° of latitude, or one-third of the full equator-to-pole distance. If Earth rotated more slowly or the pressure gradient were stronger, there might only be two cells. Under a different balance of forces, there might be four, or only one. Jupiter has five circulating cells in each hemisphere, and the balance is quite different. The Great Red Spot is a storm in one of the cells. As previously noted, Venus is rotating so slowly that it has only one cell.

FIGURE 2.8 NOAA satellite image of the equatorial band of dense cloud associated with the ITCZ

Source: NASA Earth Observatory, "The Intertropical Convergence Zone," https://earthobservatory.nasa.gov/IOTD/view.php?id=703). Courtesy of GOES Project Science Office.

The return flow in the upper atmosphere is also shown in figure 2.7, and the flow is a motion away from the equator. The Coriolis effect, therefore, causes a right-directed deflection of the return flow as well (see box 2.1). The surface flow is toward the equator, so the deflection is to the west, and the return flow deflection is to the east. This gives rise to an overall circulation pattern that is oblique to lines of latitude.

The three circulation cells are not exactly equal in size because the Coriolis effect varies in strength with latitude; it is strongest nearer the poles and near zero at the equator. In addition, the cells are not perfectly symmetrical across the equator, in part because the South Pole region is much colder than the North Pole.

The middle of the three cells, extending from 30° to 60° in both hemispheres, is referred to as the Ferrel cell. It was named for William Ferrel, who proposed its origin in 1856 and so was

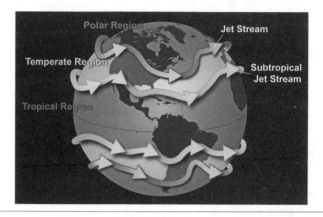

FIGURE 2.9 Location of upper atmosphere jet streams.
Source: NOAA.

able to include the Coriolis effect in his work. Its circulation pattern is opposite to that of the Hadley and polar cells. The physics explaining this pattern reversal is beyond the scope of this book.

In addition to the three circulating cells in each hemisphere, two *jet streams* (channels of very strong high-level winds) are shown at the frontal areas in figure 2.9, where two circulation cells meet within the *troposphere*. These are regions of very high pressure and temperature gradients that drive strong winds approaching two hundred miles per hour, which is stronger than typical cyclonic winds. Weather on either side of the polar jet stream can be profoundly different.

The jets have oscillatory paths around Earth that are constantly changing. When a lobe of the polar jet dips south from Canada, it can bring very cold conditions to much of North America. The polar jet is important to commercial air travel across the United States and the Atlantic. A plane flying with the advantage of the jet stream can save more than an hour of flying time.

ATMOSPHERIC CIRCULATION CELLS
AND CLIMATE ZONES

Looking back at figures 1.1 and 1.2, together with figure 2.7, we see the broad correspondence between circulation cells in the atmosphere and climate zones. Within the three-cell pattern in any hemisphere, there are two regions of upwelling—one at the equator and one at about 60° N/S, which includes traditionally rainy places such as Great Britain, northern Europe, Iceland, and Washington State in the United States, and in the south Tasmania and Terra del Fuego. The upwelling region between the polar cell and the Ferrel cell is often described as the *polar front*. It is similar to the upwelling in the ITCZ, but it is much colder.

At 30° N/S, and at the North and South Pole, the opposite occurs. These regions are at the down-welling limbs of circulation cells. The great deserts of the world are located at around 30° N/S. In these areas, rainfall in the ITCZ has removed most of the moisture from the air. As the air moves over the top of the Hadley cell and then descends at around 30°, it becomes very dry and, because air is descending, atmospheric pressure is high. Thus the deserts owe their location to the three-cell pattern of atmospheric circulation, and their dryness to loss of moisture that "rained out" in the tropics.

The uppermost surface of the circulating systems is referred to as the *tropopause*, and it is highest above the equatorial regions because the temperature at the equator is greatest and is able to drive the most vigorous upwelling.

The most temperate and fertile parts of Earth are located in the Ferrel cell. Throughout that region, average temperatures are mild, as are summer to winter temperature variations. Rainfall is fairly reliable and is approximately the same in all four seasons; however, the seasons themselves are quite distinct.

The tropopause can be up to 15 kilometers in height at the equator and less than 10 kilometers at the poles. That is one reason astronomical observations are made from the South Pole Station. Temperature and pressure gradients are much weaker above the tropopause than beneath it, and winds are very weak as well because there is so little material to move to create a wind. The height of the tropopause is hugely exaggerated in figure 2.7; drawn to correct scale, the atmospheric circulation cells would be almost invisible.

"MOTION" OF THE ITCZ AND MONSOON WEATHER

Regions affected by the ITCZ have two distinct seasons—a rainy season and a dry season—whereas temperate middle latitudes experience four distinct seasons. It is not hard to imagine that agricultural practices differ between the tropics and temperate zones. Regions under the influence of the ITCZ experience a band of heavy rainfall moving from north to south and then reversing with the seasons. In July the band of heavy rainfall is north of the equator, and in January it is south of the equator (figure 2.10).

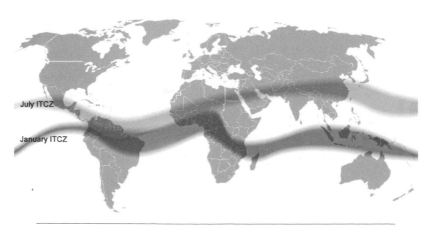

FIGURE 2.10 Apparent motion of the ITCZ from south of the equator in the northern winter to north of the equator in the northern summer.

Source: Courtesy of Mats Halldin/Wikimedia Commons.

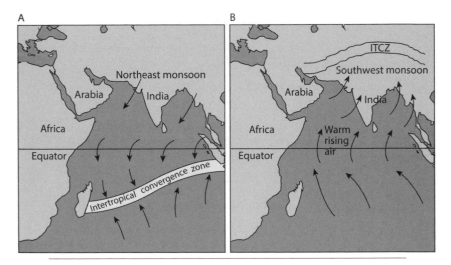

FIGURE 2.11 The association between the Indian Ocean monsoon and apparent motion of the ITCZ in (*A*) winter and (*B*) summer.

This apparent motion of the ITCZ also accounts for the Indian monsoon, an extremely important weather phenomenon Indian farmers rely on to recharge aquifers, which provide water essential for crop irrigation. The position of the ITCZ is shown at the right in figure 2.11 in summer during the monsoon and is at the left during the winter. During summer, the ITCZ is well north of its winter position, lying just south of the Himalayan Mountains. Just as described for sea breezes (see figure 2.5), the landmass retains much more heat than the ocean, causing a pressure differential that drives a landward circulation of very moist air off the ocean in the summer.

The moist air is lofted to higher, cooler elevations, where it condenses into a liquid and falls as heavy precipitation in northern India over the main agricultural areas. In winter, the

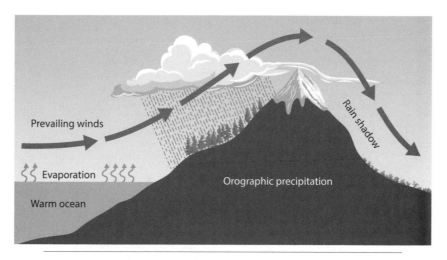

FIGURE 2.12 Orographic rainfall occurs when warm, moist air moving landward toward a coastal mountain range is forced to rise, where it condenses water vapor into precipitation. The air is then dry as it passes over the range forming a rain shadow.

ITCZ has dropped well south and the circulation has reversed, again mimicking the sea breeze reversal of circulation during nighttime hours (see figure 2.5).

The monsoon rains in India are strengthened by the *orographic effect* (figure 2.12). Moist air coming off the Indian Ocean is forced to rise over the Himalayan Mountains, and as it rises it cools and water vapor condenses, causing rainfall on the windward side of the mountains. A region of dry conditions, often referred to as a *rain shadow*, is created on the leeward side of the mountains due to this dry air. The coastal ranges in central California also exhibit this pattern of precipitation. West Africa and many other regions experience a summer monsoon as well, but the monsoons are more intense in India than elsewhere because of the dominating effect of the

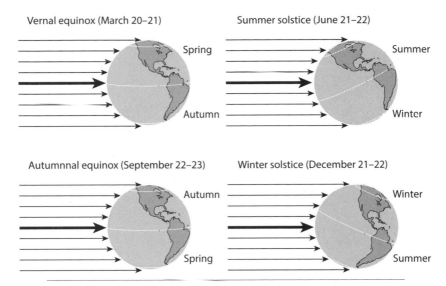

FIGURE 2.13 The apparent motion of the ITCZ to an observer on Earth is due to the Earth's tilted axis of rotation with respect to the plane of the ecliptic.

Himalayas. The orographic effect also can be seen in the way many mountaintops and islands are often cloud covered.

What makes the ITCZ move like this? In fact it doesn't move, but it appears to do so to an observer on Earth. This is another case (like the Coriolis effect) in which the frame of reference is important. The clue here is that the perceived motion is seasonal and relates to the position of Earth with respect to the Sun at different times of the year (figure 2.13).

Earth rotates on an axis that is not perfectly at right angles but is tilted about 23.5° to the plane of the ecliptic in which it rotates around the Sun (see figure 1.6). This tilt is the reason Earth experiences opposite seasons in the northern and southern hemispheres. In the northern summer, the Sun's rays warm more of Earth's northern hemisphere. The opposite is true in the northern winter, which is the southern summer. (In figure 2.13, you can see that the Sun's energy strikes Earth north of the equator on June 21.) In spring and fall, both hemispheres are heated equally. Fall and spring have similar meteorological conditions in both hemispheres, although one is transitioning from cold to warm and the other from warm to cold.

The ITCZ forms where the Sun's energy is most intense, sometimes referred to as the *climatological equator*, not at the geographic equator. The arrows in figure 2.13 point directly from the Sun toward the point on Earth where the ITCZ develops. The ITCZ can be thought of as following the Sun. It is located at the equator in spring and the fall, and is north of the equator in the northern summer and south of the equator in the northern winter. The ITCZ, in effect, remains fixed within the plane of the ecliptic. To an observer attached to the plane of the ecliptic, Earth would appear to wobble back and forth beneath a fixed ITCZ. From a fixed vantage point on Earth in the tropics, however, the ITCZ appears to oscillate north to south with the seasons.

THE ROLE OF GLOBAL OCEANS IN THE CLIMATE SYSTEM

Although north-south variations dominate the climate system, important east-west variations occur in a number of places (see figures 1.2 and 1.4). North America is climatologically divided almost down the middle: the western half is dry and the eastern half is much wetter. The dividing line where the Great Plains begins is close to the 100th meridian west. South America shows similar variations east to west at a different longitude.

The key to understanding what causes these variations is to include the effect of ocean circulation. The pattern of surface water movements is shown in figure 2.14. It is important to emphasize that this is the motion of water at the surface. At depth, waters may flow in the opposite direction to water at the surface,

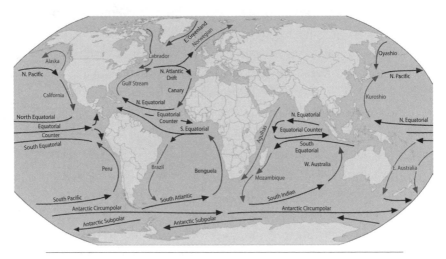

FIGURE 2.14 Simplified pattern of surface ocean currents.

Source: Courtesy of Michael Pidwirny/Wikimedia Commons.

primarily because surface water flow is strongly influenced by the global pattern of winds. This is somewhat analogous to circulation in the atmosphere, although the oceans are much denser and are less subject to the Coriolis effect.

Heat in the surface ocean is easily imparted to the air above it and is, in effect, carried along with the water as it slowly moves. Water, having once gained warmth, releases heat much more slowly than air, and thus warm air can be carried into regions where one might expect much cooler air from atmospheric circulation alone. The reverse is also true: cooler water can bring cool air toward the equator.

Ocean currents respond to the Coriolis effect and are driven by pressure differences similar to those of atmospheric motion. But oceans are much denser and therefore are less subject to the Coriolis effect, and there is no counterpart to the three-cell system of atmospheric circulation.

In addition, the density of water masses is governed by both temperature and salinity of the ocean water and circulation in the oceans, which is referred to as *thermohaline*. Of course, ocean currents are profoundly influenced by landmasses that deflect them from where they might have traveled had the landmass not been present. This is not strictly the oceanographic equivalent of orographic effects on rainfall described previously, but it does bear some similarities. The only place on Earth where currents flow unimpeded by landmasses is in the Antarctic Circumpolar Current, which, as its name implies, sweeps around the entire continent of Antarctica without intersecting any obstacles.

Surface water motion is characterized by circulating *gyres*, which are much larger in scale and move much more slowly than atmospheric circulation patterns. Horizontal and vertical friction of water against the continents and seafloor also play an important role. The North Atlantic Gyre is an almost circular feature with a clockwise rotation that hugs the coast of both North America and northwest Africa and crosses the central equatorial Atlantic much like the trade winds, where the waters are warmed in a region of high solar irradiation. Traveling west, they meet the landmass of North America and are deflected north along the east coast of the continent, carrying warm water northward. The water warms the atmosphere, and air becomes both warm and moist, which leads to additional warmth and precipitation along the east coast of the United States and further north into the North Atlantic.

The west coast of North America is influenced by a gyre in the Pacific that has exactly the opposite effect on the western regions of the continent, where cool waters from the North Pacific flow south along the west coast. This means that cool ocean currents influence the western coastal regions of the United States as warm currents influence the east coast. The current carrying warm waters north along the U.S. east coast creates relatively

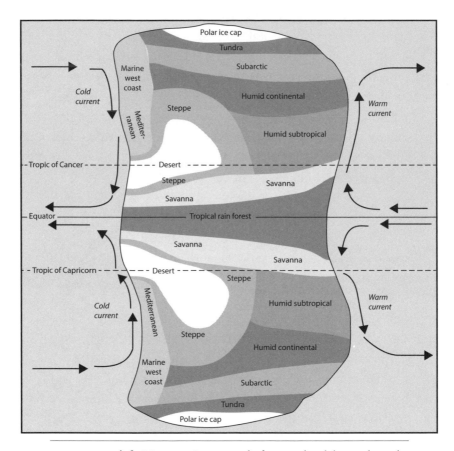

FIGURE 2.15 A fictitious continent stretched across the globe north south showing the influence of warm ocean currents on the east coast climate zones and cold currents on western regions.

warm moist conditions there. In the west, cooler water transported south produces dryer air conditions. The net effect is to distort the strong north-south variations so that wet temperate regions expand in the east and dry conditions expand in the west. (In chapter 6, you will see that these ocean conditions influence the pattern of hurricane tracks as well.) A rendition of the plan is shown in figure 2.15 for a fictitious continent that stretches from pole to pole and contains all the basic climate zones, but overall it represents a fair approximation of the Americas.

An additional factor is that the west coast of the United States experiences strong orographic rain effects, whereas the east coast—where mountains are much older, eroded, and lower than in the west—does not experience significant orographic effects. The same is true for the west and east coasts of Latin America. In both cases, these effects contribute considerably to the east-west variation in climate conditions.

3

CLIMATE DYNAMICS

Natural Variations

THE climate we experience today is not at all like Earth's climate only twelve thousand years ago. At that time, Earth was still in an ice age, or *glacial period*, and ice sheets like the one now covering Greenland extended as far south as New York and covered most of Europe. Earth's average temperature was perhaps 10°C (17°F) colder than today, so the average Earth temperature would have been 5°C (40°F). That is about average for Canada today. At the height of the last ice age, those parts of Earth that could be described as habitable would have shrunk to a small band of land in the equatorial regions.

During this time, the polar cells would have been huge, and the Hadley and Ferrel cells would have contracted. Our second approximation provides no insight into the cause of these kinds of change. Chapter 3 describes a third refinement to the emerging picture of the climate system, adding a basic treatment of natural changes in climate—that is, climate changes not brought about by human activity.

Warm periods between glacial periods are called *interglacial periods*, and Earth sustained an average temperature much higher than today in the Turonian period from 93.5 to 89.3 million years ago. During that time, temperatures were probably as much as 5°C warmer than today. During the Eocene epoch, 56 to 33.9 million

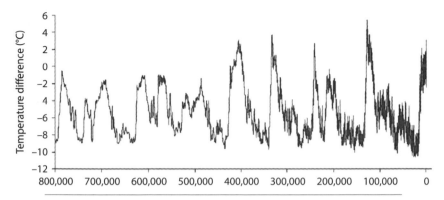

FIGURE 3.1 Temperature record to 800,000 years before present derived from ice core data. Temperature is given as departure from the present (0°C is the present). Many such renditions of this record are available.

Source: Jean Jouzel, Valérie Masson-Delmotte, Olivier Cattani, et al., "Orbital and Millennial Antarctic Climate Variability Over the Past 800,000 Years," *Science* 317, no. 5839 (2007): 793–796.

years before today, temperatures may have been higher still. These very warm conditions are unusual in Earth's history. The temperature record in figure 3.1 is based on an analysis of ice core data representing the last 800,000 years. These data show that Earth has usually been much colder than it is today. Today's temperature is typical of the temperature of most interglacial periods.

The most recent period in which Earth's temperature was about the same as it is today was around 120,000 years ago, during the relatively brief *Eemian* interglacial period. Interglacial periods are, in fact, typically quite brief compared with the length of glacial periods. Over the last million years or so, glacial periods (ice ages) and interglacial periods have alternated rhythmically, with warm periods occurring every 100,000 years or so. Interglacial periods make up barely 5 percent of the record of the last 400,000 years. As our species evolved during the last million years, most of the time the northern hemisphere looked like the image in figure 3.2.

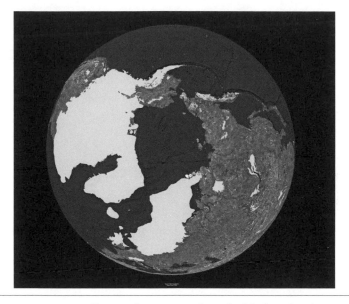

FIGURE 3.2 Extent of ice cover at the end of the last glacial maximum (LGM).

Source: "Maximum Ice Extent at the Last Glacial Maximum," Quarternary Palaeoenvironments Group, University of Cambridge, February 7, 2012, https://www.qpg.geog.cam.ac.uk/lgmextent.html.

LONG-TERM CLIMATE CYCLES: ORIGIN OF GLACIAL AND INTERGLACIAL PERIODS

Since the earliest days of development of the geological sciences, it has been recognized that some processes of Earth dynamics must be extremely rhythmic. Sedimentary sections show remarkably repetitious banding, sometimes referred to as *cyclotherms* (figure 3.3).

The source of the cyclical banding was not understood until a controversy began in the 1920s when Milutin Milankovitch, a Serbian mathematician and astronomer (among other pursuits), proposed that the cycles were caused by variations in the orbit of

FIGURE 3.3 Rhythmic banding of layers in a sedimentary rock formation.
Source: Courtesy of Verisimilus/Wikimedia Commons.

Earth around the Sun, together with other aspects of Earth's rotational dynamics. Geologists had not considered such a mechanism, believing that those changes would be too small to cause the strong banding seen in rock sections. Furthermore, many sedimentary sections (most, in fact) did not show any form of repetitious banding.

The Milankovitch mechanism was not generally accepted until the mid-1970s when core samples taken from beneath the deep seafloor where sediments were deposited slowly in very still conditions, without any postdepositional disturbance, could be examined and dated by fossil evidence, and later by isotope methods.[1] The fossil and isotope information also provided evidence of warm and cool cycles in Earth's history that exhibited distinct periodicities. Vast areas of the deep oceans have conditions where sediments are deposited in this way, and systematic sampling of these sediments, primarily by research vessels from the Lamont-Doherty Earth Observatory, showed that periodicities of sedimentary deposits coincided closely with those of the orbital motions described by Milankovitch.

To understand how this came about, we need to consider aspects of Earth's rotational dynamics that are almost imperceptible to us. As well as spinning on its axis once a day (this is how a day is defined), and revolving around the Sun once a year (the definition of a year), the Sun/Earth system executes three other motions that vary so slowly that astronomical observations are needed to identify them (figure 3.4).

Earth's orbit around the Sun is in the form of an ellipse (although it is nearly circular; see figure 3.4) with the Sun at one focus. The shape of the ellipse changes very slowly, becoming more circular and then more distended over a period of 100,000 years.

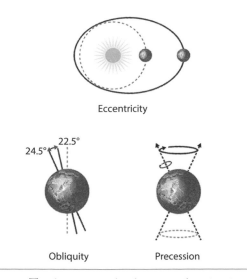

FIGURE 3.4 The three main orbital motions that give rise to Milankovitch cycles.

Source: Japan Agency for Marine-Earth Science and Technology (JAMSTEC), "How Does Orbital Variations Affect Ancient Antarctica? East Antarctic Ice Volume Change During the Pliocene and Early Pleistocene," October 27, 2014, http://www.jamstec.go.jp/e/about/press_release/20141027/.

Eccentricity is the amount Earth's orbit differs from that of a circle. The ellipse itself slowly rotates with the Sun as the center of rotation, but this has no effect on climate. At the same time, the tilt of Earth's axis with respect to the plane of the ecliptic changes, becoming a little more upright, and then more tilted—this is known as *obliquity*—with a full cycle taking about 41,000 years. Finally, the orientation of the spin axis executes a slow rotation, which looks something like the way a spinning top behaves as it is slowing down. A full cycle of that motion takes around 23,000 years and is called *precession*. These periods of planetary motion coincide exactly with the periodicity of warm and cool periods seen in deep-sea sediment cores. The cycles, especially the 100,000-year eccentricity cycle, are particularly well observed in ice core records (see figure 3.1).

All planets in the solar system, as well as the Sun itself, exhibit these motions, and some are much more exaggerated than Earth's. For instance, Mars experiences obliquity changes from at least 15° to 35°, and possibly greater. Uranus has an obliquity of more than 90°, meaning that its axis of rotation is directed toward the Sun. Venus is very slowly rotating on its axis in the opposite direction to all other planets, most likely because it has flipped a full 180° due to a very strong obliquity motion. Earth's obliquity shows unusually small variation because the gravitational effect of our moon—the largest satellite of any planet in the solar system relative to the size of the planet itself—constrains the motion.[2]

The temperature changes seen in ice cores are much sharper than can be attributed to Milankovitch's mechanism alone, although timing of maxima and minima are in the correct positions. The orbital variations are slow and rhythmic, but the record includes very sharp peaks and troughs. The sharpening of the peaks is due to *feedback* effects. Feedback effects (box 3.1) are extremely important in climate science and are one of the main reasons for uncertainty in climate predictions, as you will see later.

BOX 3.1 FEEDBACK EFFECTS

As used in science (not specifically climate science), the term *feedback* describes the way in which part or all of the output can be returned to the input in a process shown stylistically here. Feedback is commonly used in electric circuit theory to indicate an amplifier.

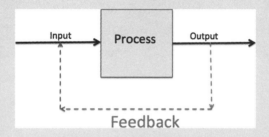

BOX FIGURE 3.1.1 Feedbacks demonstrated via electric circuit theory.

If the portion of output is added back to the input, the feedback is described as positive; if it is subtracted, it is said to be negative. Positive feedback gives rise to enhanced output, and negative feedback suppresses output.

A familiar example is noise at a restaurant. If you are the first to sit down to lunch with a friend, you can usually speak in a normal voice and be heard, but as others sit down and begin to talk, you need to raise your voice to be heard. This can go on until you just about need to shout at one another—noise creates even more noise, so the feedback is termed positive. When you turn up the volume on a radio, you are using positive feedback in the electronics.

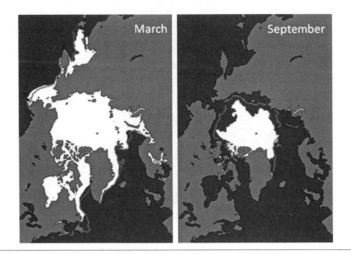

FIGURE 3.5 Typical extent of Arctic sea ice in the northern winter (*left*) and summer (*right*).

Source: U.S. National Snow and Ice Data Center (NSIDC).

In the present case, feedback operates in the following ways: the orbital variations begin to trigger a glacial period when the three orbital motions of the Milankovitch cycle coincide in such a way that northern winters are especially cold. Remember that the tilt of Earth's rotational axis explains our seasons, so any change in tilt will affect seasonality. This has a significant effect on the ice cover in the northern hemisphere (less so in the southern hemisphere). Ice cover, and particularly sea ice in the Arctic, shrinks in extent during the northern summer and grows in extent in the northern winter (figure 3.5) in a cycle that is repeated annually.

The maximum ice extent varies from year to year, as does the extent of the summer retreat. This seasonal advance and retreat is quite natural, and it has been observed for centuries. When there is a particularly large advance, the Denmark Straits between Greenland and Iceland can become completely choked with ice, and in the nineteenth century several expeditions crossed the frozen sea from Iceland to Greenland and returned. In times of unusually large retreats, the Northwest Passage between the Pacific and Atlantic oceans can become ice-free and is navigable for several weeks in the summer. One concern today thought to be a clear expression of climate change is a systematic decrease in summer ice minimum over at least the last twenty years. The North Pole is often ice-free in summer months today, suggesting that the region is experiencing a long-term warming.

When changes in orbital parameters in combination favor the development of especially cold northern winters, the natural seasonal ice advance in winter becomes greater as well. Feedback comes into play here because ice is highly reflective (albedo of about 0.35; see figure 1.12). As the ice advance grows larger, Earth's total albedo increases (for example, the albedo term in the equation in box 1.3 that was 0.3 becomes larger, maybe 0.4). Since more solar energy is being reflected, less is being received to warm Earth; hence the critical balance of incoming and outgoing solar energy is disturbed, and Earth cools down to achieve a rebalancing. This is described as a positive feedback—one effect reinforcing another (box 3.1).[3] In this case, the Milankovitch forcing gives rise to cooler northern winters, which become even cooler through the feedback effect. The process is referred to as the *ice albedo feedback* effect, and it is diagramed in figure 3.6.

The opposite effect happens when Earth comes out of a glacial period. Beginning with the setting depicted at the top of figure 3.6, the ice sheet is large and very little land surface is exposed, similar to how Greenland appears today. As the ice sheet begins to melt, more of the land surface becomes exposed, and the total albedo is lowered. When more land surface is exposed, more energy arriving from the Sun is available, so Earth warms more than it would by Milankovitch forcing alone. The lower part of the illustration

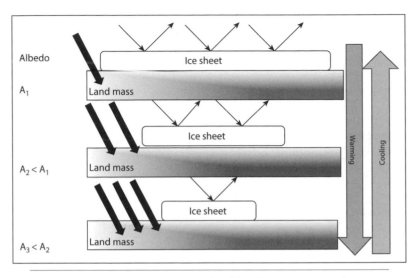

FIGURE 3.6 The ice albedo feedback effect. The arrows represent solar energy.

shows the ice sheet at maximum retreat, with a large amount of land surface exposed.

A second feedback is also involved. Carbon dioxide is readily absorbed in seawater. In fact, the ocean is Earth's largest reservoir of carbon (not the atmosphere, biosphere, or rock formations). Absorption of CO_2 into the ocean has no effect on ocean temperature, but the rate of absorption is temperature dependent—colder water absorbs more of the gas.[4] As the ocean begins to cool, the CO_2 content of the atmosphere drops due to enhanced ocean absorption. With less of this greenhouse gas in the atmosphere, a further cooling will occur. This too is a positive feedback because it enhances the primary Milankovitch-induced cooling, so both the ice albedo feedback and the ocean absorption feedback are positive by definition (see box 3.1).

The ice albedo feedback effect can accelerate warming and cooling, and this largely explains the jaggedness of the ice-core record.

That said, there are conundrums in the orbital-variation story. All calculations of the Milankovitch mechanism would say that the strongest signal should be the 21,000-year obliquity cycle, and the smallest ought to be the 100,000-year eccentricity cycle. Yet what we observe is the opposite; the peaks at 100,000 years are the most prominent. Current thinking is that the obliquity signal might be countered and diminished due to its opposing effects in the two hemispheres—increased ice sheets in the north come at a time of reduced ice sheets in the south. That would certainly blunt the obliquity signal.

The strength of the 100,000-year eccentricity signal is harder to explain because no feedback mechanism of the type described here operates on it. One possibility is that eccentricity and obliquity may be phase-locked, meaning that the two signals may peak at the same moment and in this way amplify the eccentricity signal. It should be noted that the dominance of the 100,000-year signal is relatively recent, having come into prominence only in the last five million years of Earth history. An explanation for this is not known.

A further puzzle is the jagged asymmetry of the temperature record (see figure 3.1). Earth descends fairly slowly into a glacial period but comes out of one into an interglacial period much more rapidly. This is particularly apparent in the most recent 100,000 years of the record. This creates an asymmetric, sawtooth appearance in the temperature record that has yet to be explained.

The climate system includes scores of feedback effects, such as between ocean and atmosphere, atmosphere and biosphere, *cryosphere* and ocean, etc. These effects can act in either direction—to enhance or to counter primary climate forcing—so both positive and negative feedback effects operate in the climate system. These effects also act on many different time scales; some are virtually instantaneous, others are very slow.

NATURAL CLIMATE VARIATIONS ON TIME SCALES OF YEARS TO DECADES

The dominant signal of climate variation experienced any place on Earth is the change of seasons, even in Hawaii or at the South Pole where conditions don't change greatly from season to season. Seasonal variations are most prominent in middle latitude, temperate zones that typically experience four quite distinct seasons of almost equal length. In very high and very low latitudes, seasons are less distinct—many parts of the tropics have a dry and a wet season but experience relatively little change in temperature between the two because the incidence angle of the Sun is high at all times of the year. The annual range of temperatures in New York, for instance, is 4 times greater than in Hawaii. The high latitudes also experience somewhat less distinct seasons due to the low Sun angle, but all parts of Earth experience four seasons, although they may be defined differently.

El Niño Southern Oscillation

One expression of natural climate variability is the *El Niño Southern Oscillation* (ENSO). Although it was most prominently developed for the tropical Pacific, ENSO has influences in many other parts of the world. El Niño ("the boy" in Spanish) was named for the Christ Child because it disrupts normal weather conditions around Christmastime on the west coast of Latin America. The other ENSO phase is La Niña ("the girl" in Spanish). El Niño is often referred to as the warm phase and La Niña as the cold phase of ENSO. Either phase of ENSO can simultaneously give rise to drought conditions in one part of the world and heavy rains in another.

Understanding ENSO begins by understanding what is meant by "normal conditions"[5] in the tropical Pacific and the forces that can cause disturbance to those normal systems. El Niño and La Niña are departures from the norm. Until now I have focused on the first-order expressions of climate that are developed most strongly in a north-south orientation. The processes driving ENSO, in contrast, are developed as an east-west circulation across the tropical Pacific.

In conditions described as normal, trade winds blow across the Pacific from east to west (figure 3.7; see also figures 2.8 and 2.9). The ocean surface waters in the equatorial Pacific are very warm because of the high Sun angle (see figures 1.10 and 1.11). This creates a warm surface layer with fairly uniform water temperature

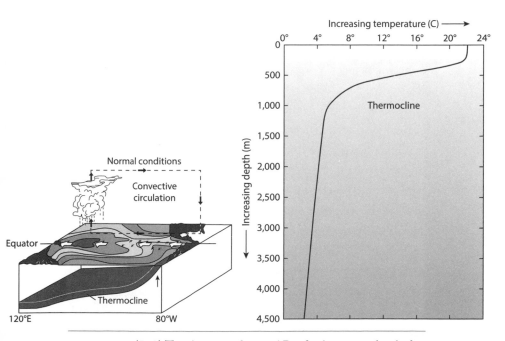

FIGURE 3.7 (*Left*) The elements of normal Pacific climate are sketched. (*Right*) A typical ocean temperature profile. Note the rapid drop in temperature between 500 and 1,000 meters. This region is referred to as the thermocline.

down to a few tens of meters below the surface (in oceanography this is known as the mixed layer), after which temperature decreases sharply. At a thousand meters or so below the surface, water temperature again becomes more uniform and decreases slowly with depth (see figure 3.7). The region of rapid decrease in temperature between the shallow and deep regions is known as the *thermocline*. Below the thermocline ocean temperatures are affected very little by processes at the sea surface. The depth and gradient of the thermocline is considerably different between tropical and colder oceans and changes with the seasons, but the three-part structure is usually present.[6] In the equatorial Pacific, under normal conditions, the thermocline is much deeper in the western Pacific than in the east. Warmer water is found to a greater depth adjacent to the coast of Peru where the thermocline shoals are near the surface. This shallowing leads to an upwelling of nutrient-rich cold waters along the west coast of Latin America, which supports highly productive coastal fisheries during normal conditions.

Several factors interact to create the system of atmospheric circulation that characterizes these normal conditions. The trade winds literally push warm surface waters to the west, creating a pool of warm water in the Indonesian region. At the surface, the warm pool of water heats the atmosphere and creates a low pressure system, and the cooler water in the east creates a relatively high pressure cell. This sets up an east-west circulation system similar to the north-south circulation associated with the Hadley cells, or the sea breeze system (see figure 2.5). The low pressure region in the west forms a zone of rising warm moist air and associated precipitation.

The circulation pattern across the central Pacific in normal conditions is known as the *Walker circulation* after Sir Gilbert Walker, a British applied mathematician assigned to India in 1904 as the director-general of observatories to develop an understanding of *monsoon* failures, which lead to drought and devastating famine. Massive famines in 1899 caused millions of deaths. He spent painstaking years sorting through large amounts of meteorological data in hopes of understanding the cause of failure of the Indian monsoon (see figure 2.11). He had access to a vast collection of meteorological observations from the British Empire and recognized that high pressure systems developed in the western Pacific, centered over Darwin, Australia, were always matched by low pressure systems in the east, centered near Tahiti (see figure 3.7). The reverse was also true. If a low pressure system formed in the west, it would be associated with high pressure in the east. He found no instance in which the pressure was simultaneously high or low at both locations. This is referred to as the Southern Oscillation because the centers of the high and low pressure systems lie south of the equator and vary arythmically.

To quantify this observation, Walker developed the Southern Oscillation Index (SOI), which measures the difference in sea level pressure between Tahiti and Darwin. The formula is a bit more complicated than a simple difference and reads as follows:

$$SOI = 10 \frac{(P_{diff} - P_{diffav})}{SD(P_{diff})}.$$

The term in the numerator is the pressure difference of a current month minus the average pressure difference, and the term in the denominator is the standard deviation of the pressure differences for that month. This yields a dimensionless index with values ranging from about −35 to 35.

The SOI oscillates between these values in a quasi-periodic manner. It is not as reliably rhythmic as the seasons, but it is not entirely random either. Technically, it is a low-order, nonlinear oceanic oscillator constrained in an equatorial waveguide, meaning that it is confined to the equatorial region and has a quasi-chaotic behavior. The important result of Walker's statistical analysis was that the SOI correlated with rainfall events in the Indian Ocean. Walker did not know the cause of the oscillation, or why it correlated with rainfall, but he glimpsed the beginning of modern ideas about climate prediction. If the SOI could be forecast, then rainfall (and especially the critically important Indian monsoon) could be predicted as a derivative consequence. Most critical, absence of the monsoon, which leads to drought conditions, could be anticipated and systems set in place to mitigate the consequences.

To understand the climate pattern in the Pacific during El Niño, imagine that the trade winds slacken considerably for unspecified reasons, or even reverse, as shown by the direction of the arrows in figure 3.8. As a consequence, the pool of warm water normally pushed to the west by the trade winds slides east toward the central Pacific, bringing with it low pressure and upwelling atmospheric convection, with its associated rain normally found in the Indonesian region. The Walker circulation

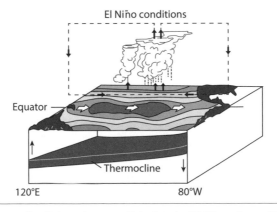

FIGURE 3.8 Conditions that prevail during the El Niño phase of ENSO. Note the flattening of the thermocline and reversal of the trade winds.

may be completely broken or reversed, and the thermocline also loses its upward slope to the east. Rain now characterizes the central and eastern Pacific. As a consequence of the flattened thermocline, upwelling of cold, nutrient-rich water off the coast of Peru is interrupted, critical fish populations are displaced, and the local fishing industry suffers.

The opposite phase, La Niña (the cold phase of the ENSO), can be thought of as an exaggeration of normal conditions (figure 3.9). Trade winds blow more strongly than normal, a warm pool in the west piles up even more than under normal conditions, and the thermocline is so steep that only a very shallow layer of warm water is near the surface in the eastern Pacific, in the fishing grounds. The Walker circulation is not broken and may be intensified, and upwelling can become quite intense.

What is generally termed *weather* is atmospheric phenomena—wind, rain, sunshine, etc. Climate can be thought of as the long-term average of weather, but as the mechanism of ENSO demonstrates, the oceans are critically involved in creating those

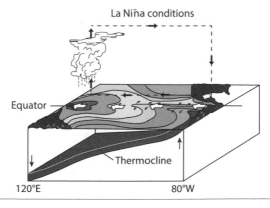

FIGURE 3.9 Conditions that prevail during the La Niña phase of ENSO.

climate conditions. And recall from chapter 2 that the oceans provide a strong influence on east-west variations in the global climate system. Two very different ocean conditions—El Niño and La Niña—impart changes to the overlying atmosphere, although the direction of causation is not straightforward. Warm ocean water certainly heats air above it, causing lower atmospheric pressure, and the wind field produced by pressure gradients in the atmosphere causes movement in the water masses, pushing warm water west under normal conditions. Ocean and atmospheric processes are coupled and respond to one another: any change in one will lead to a change in the other, which may then cause a change to the first. This is a complex set of ocean-atmosphere feedback effects—interrelationships first described by Norwegian scientist Jacob Bjerknes—and they are often referred to as *Bjerknes feedback effects*.[7]

Dynamic coupling between ocean and atmosphere is demonstrated in figure 3.10, which compares the SOI, based on atmospheric pressure conditions, with sea surface temperature (SST) anomalies of the ocean in a box centered on the equator and spanning 170° W to 120° W (described as *NINO3.4*). The two signals are in almost perfect antiphase (note that the scale for the SST figures has negative values upward). When the ocean temperature is high, SOI has low values, and the opposite effect (low ocean temperature and high SOI) is just as prominent. The strong positive peaks in ocean temperature correspond to El Niño phases, and the strong negatives correspond to La Niña phases.

Sea surface temperatures, salinity, surface winds, and the temperature profile below the surface are now routinely recorded on arrays of buoys moored across the equatorial Pacific (figure 3.11). These data are continuously monitor and sent via satellite to central facilities where they are used in ENSO forecasting schemes, which I describe in the next section.

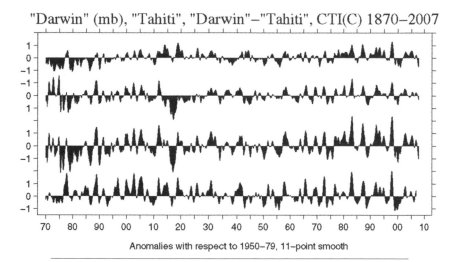

"Darwin" (mb), "Tahiti", "Darwin"–"Tahiti", CTI(C) 1870–2007

Anomalies with respect to 1950–79, 11–point smooth

FIGURE 3.10 Comparison of sea surface temperature anomalies (SST) and the Southern Ocean Index (SOI) that illustrates their relationship. Top row is Darwin SLP, second line is Tahiti SLP, third is Darwin minus Tahiti, and the fourth is SST.

Source: Joint Institute for the Study of the Atmosphere and Ocean, "Southern Oscillation Index Derived from Marine Observations, 1800*—March 2008," http://research.jisao.washington.edu/data/soicoads2/.

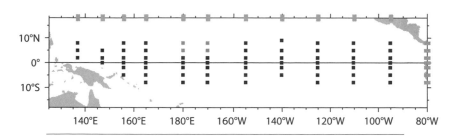

FIGURE 3.11 The TAO (Tropical Ocean Atmosphere) array of measuring buoys that sense conditions required to forecast ENSO. ATLAS buoys have instruments that measure conditions below the surface in addition to surface conditions.

Source: NOAA Earth System Research Laboratory, https://www.esrl.noaa.gov /psd/enso/rapid_response/data_pub/.

Global Impact of ENSO Variations

Although ENSO is generated in the tropical Pacific, both warm and cold phases have distinct impacts well outside the equatorial Pacific itself (figure 3.12). This global pattern is generally referred to as ENSO *teleconnections.* These connections arise in two ways.

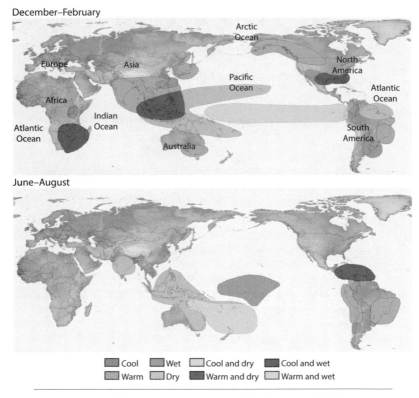

FIGURE 3.12 The two panels show the global impact of El Niño via teleconnections in two seasons. Note that although El Niño is most strongly developed in the Christmas period, its effect can be experienced well into the year.

Source: NOAA.

First, the Walker circulation is a three-dimensional flow; at the equator, it develops a strong east-west orientation, but the circulation extends into the north and south Pacific oceans. The anvil-shaped system that dominates in the upper panel of figure 3.12 is an expression of the three-dimensionality of the Walker circulation.

Second, the strength and extent of the *SST anomaly* are both quite significant, and the global circulation system of the atmosphere previously described readily moves the tropical disturbance toward the pole. Because rainfall shifts to the mid-Pacific or further east during El Niño years, the Indonesian and Australian regions become unusually dry (figure 3.13). India is warmer, and as Walker observed, the monsoon rains often fail. Other parts of the world, such as the southwest coast of the United States and the Peruvian coast of Latin America, can be unusually wet during El Niño years. Although El Niño is associated with warming in the central equatorial Pacific Ocean, the global atmospheric circulation patterns ensure that its effects are transported throughout the world in a nonuniform manner. El Niño is as often associated with unusually wet conditions or unusually dry conditions, depending on the shape of the teleconnections and the time of year.

This is the case for La Niña (the cold phase) as well. It can bring equally strong departures from normal conditions, both wet and dry. Flooding in Queensland (northeast Australia) in 2010–11 and in Pakistan occurred during La Niña years. Although El Niño is the most discussed ENSO phase, La Niña conditions can bring far worse weather conditions to some parts of the world.

ENSO has an important global expression, but its impacts are considerably muted at greater distances from the Pacific. Europe, for instance, experiences essentially no effects from ENSO. It is far more influenced by the North Atlantic Oscillation (NAO). In the United States, the primary effect is felt on the west coast, where

El Niño

El Niño conditions in the tropical Pacific are known to shift rainfall patterns in many different parts of the world. Although they vary somewhat from one El Niño to the next, the strongest shifts remain fairly consistent in the regions and seasons shown on the map below.

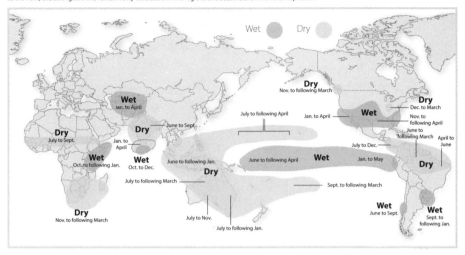

La Niña

La Niña conditions in the tropical Pacific are known to shift rainfall patterns in many different parts of the world. Although they vary somewhat from one La Niña to the next, the strongest shifts remain fairly consistent in the regions and seasons shown on the map below.

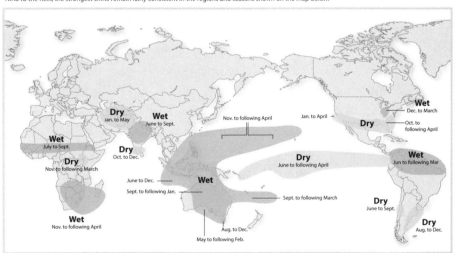

FIGURE 3.13 The effect of El Niño and La Niña on rainfall patterns.

Source: International Research Institute for Climate and Society.

El Niño can bring very strong precipitation anomalies to southern California. La Niña conditions on the west coast generally do not bring very noticeable changes to normal conditions. Neither El Niño nor La Niña has noticeable effects on the east coast.

ENSO does, however, modulate the number of landfalling Atlantic hurricanes. El Niño years are associated with significantly fewer hurricanes, and La Niña years are associated with a greater number. This may seem counterintuitive because hurricanes develop in the warmer months of the year in the tropics and are fueled by warm ocean water, but they also require relatively mild upper-level atmospheric winds. One consequence of El Niño's global impact is that upper-level winds become relatively strong, increasing vertical *wind shear,* which inhibits the formation of hurricanes. The net effect is that hurricane activity is suppressed in the Atlantic in El Niño years despite it being the warm phase of ENSO (figure 3.14).

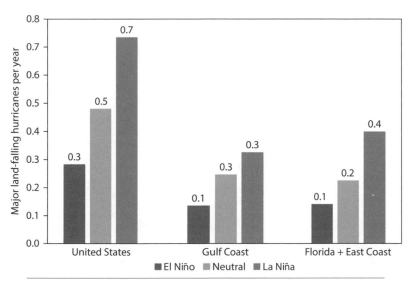

FIGURE 3.14 Mean annual land-falling major hurricanes in the United States, 1900–2016.

(The effect of global temperature increase on the strength and pattern of hurricanes is discussed in chapter 6.)

ENSO PREDICTION: METHOD AND APPLICATIONS

One important aspect of any weather forecast is that it is not precise, it is probabilistic. It gives the probability, or likelihood, of particular weather events occurring at an approximate time in the future. The following NOAA Weather Service forecast was for Piermont, New York, a few years ago:

> Tonight: A chance of rain after 1 am. Mostly cloudy, with a low around 67. South wind between 3 and 6 mph. Chance of precipitation is 30 percent.
>
> Sunday: Rain and possibly a thunderstorm. High near 77. South wind between 5 and 7 mph. Chance of precipitation is 80 percent. New rainfall amounts between three quarters and one inch possible.
>
> Sunday Night: Rain. Low around 68. South wind around 6 mph. Chance of precipitation is 80 percent. New rainfall amounts between three quarters and one inch possible.
>
> Monday: Rain. High near 79. Southeast wind around 6 mph. Chance of precipitation is 80 percent. New rainfall amounts between a tenth and quarter of an inch possible.
>
> Monday Night: A chance of showers. Mostly cloudy, with a low around 66. Chance of precipitation is 50 percent. New rainfall amounts of less than a tenth of an inch possible.

Most of us would read that as it will probably rain on Sunday, but maybe not. The word "chance" appears seven times. The chance of rain is given as a percentage, the temperature prediction is stated as being "about" or "between" or "less than" an expected value, and rainfall amounts are given as ranges. Predictions change and become more reliable as the target time of the prediction gets closer. The forecast for Sunday available the previous Monday suggested a 30 percent change of rain, and that percentage crept up through the week until it reached 80 percent. That may seem obvious—the closer you are to the time of interest, the better the prediction should be. But it is an important idea for ENSO predictions because both El Niño and La Niña have their peak at a fixed time of year. That means that part of the prediction task is constrained because the time of the year when El Niño is at its peak does not vary with every new occurrence. Similar to weather predictions, ENSO predictions become more accurate as the time of the peak is approached, but they nevertheless remain probabilistic.

Method

Estimates of future conditions are based on computer models, as are climate change predictions. The International Research Institute for Climate and Society (IRI) and other institutions employ a multimodel approach to forecast the phase of ENSO. The IRI assesses twenty-four different models in making its forecast. Sixteen of the models are *dynamical models*, and eight are *statistical models*. The prediction for each of the different models is shown in figure 3.15.

The diagram is read with time running from left to right. The first two black dots on the left are observations (OBS), and the other lines trace forecasts for each of sixteen models. These models predict the SST anomaly in the NINO3.4 box in the Pacific (scale on the left side). Atmospheric conditions are then determined from the predicted SSTs. This is a two-stage process in which ocean temperature is used as a boundary condition in the calculation of critical atmospheric factors such as rainfall and winds. In that sense,

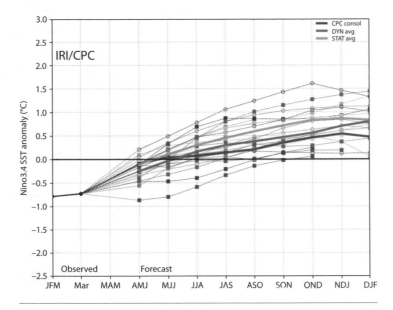

FIGURE 3.15 A typical form of prediction graphic showing forecast SSTs made by a suite of dynamic and statistical models. This graph shows the model predictions of ENSO from April 2018 to early 2019.

Source: International Research Institute for Climate and Society.

the models are not predicting rainfall per se; they are predicting changes in the ocean from which changes in the atmosphere are deduced and from which rainfall can be calculated. This two-stage process is necessary because the time scale of changing patterns in the oceans is slow compared with that of the atmosphere.

One way to appreciate this is with a simple thought experiment. Image your bathroom air is at 15°C (60°F), and you fill the bathtub with water at 30°C (86°F). The air in the room will reach about the same temperature as the bath water quite quickly. But if you do the reverse, starting with a tub of water at 15°C (60°F) and heating the air in the room to 30°C (86°F), it will take a very long time for the water to reach the air temperature (the water will eventually reach the air temperature if the heat can be maintained in the room).

Because natural processes in the ocean operate on time scales of weeks to months, predications of ocean conditions can be made weeks to months into the future. Although various models differ quite a bit in their forecasts, they all show relatively slow changes, just a few tenths of a degree Celsius per month. Nevertheless, the model forecasts shown in figure 3.15 don't look very encouraging. They begin with negative SST anomalies associated with the remnants of the previous La Niña, but as many predict neutral conditions as predict La Niña or El Niño conditions. Dynamic models are no better than statistical models in this regard.

The difference between these two types of modeling approaches is significant. Dynamic models use mathematical expressions to represent the physical laws that describe ocean and atmospheric processes. These models are initialized with current conditions in the ocean and the atmosphere, and the equations governing ocean processes are coded into a computer and used to determine future conditions six or more months ahead. The computer simulation runs the equations into future times. Statistical models (as the name implies) analyze the statistical properties of past observations of ocean and atmospheric conditions as far back as thirty to fifty years and use those to project forward. The idea is to identify typical precursors to changes in SST anomalies associated with ENSO. Statistical models do not use the equations of ocean and atmospheric physics at all. Many different statistical tools can be applied, and these give rise to eight of the models used in the IRI forecasts.

This spread in predictions is typical of climate forecasts and will be encountered again in the discussion of global climate change. It results from two factors. First, there is no agreement on a single "correct" model representation of climate physics. It is not possible to precisely represent the physics of the climate system in computer code, and a variety of different approximations techniques are used. Second, it is not possible to perfectly describe Earth's properties in all their details. But even more fundamentally, the mechanism of ENSO involves feedback between ocean and atmosphere that operate at two very different time scales, and this ensures that underlying *chaotic properties* inherently limit the ability to predict more than a short time into the future.

A different form in which an ENSO forecast is given is shown in figure 3.16a. This forecast was made in July 2011 with probabilities (the vertical axis) given in three-month units to April-May-June (AMJ) of 2012 (the three-letter sequence on the horizontal axis refers to the three months). For December-January-February (DJF), when El Niño would peak, the graph is read to say that there is a 14 percent chance of El Niño (the bar on the left), a 26 percent chance of La Niña (the bar on the right), and a 60 percent chance of a neutral condition. The forecast at that time is therefore predicting that the greatest likelihood is that conditions will be neutral, and there will be neither El Niño nor La Niña conditions in December-January-February.

A very different picture pertained in 2009, for instance, when the forecast in May gave high probabilities of an El Niño event in December that year, and that turned out to be the case (figure 3.16b). As early as JJA (June-July-August), there were strong indications that El Niño would develop. No doubt weather forecasters could use a similar scheme, and some of the graphics available on the Weather Channel and Weather Underground convey similar information, always in terms of probability or likelihood.

ENSO Predictions for Sustainable Development

The burden on economic development created by infectious disease, particularly in poor countries in tropical regions, has been widely discussed. People who are sick are far less productive than those

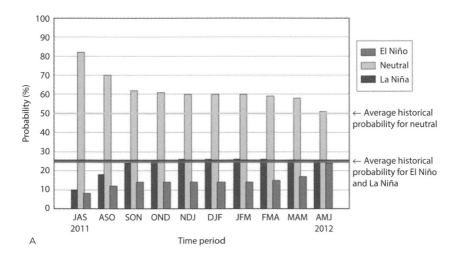

A

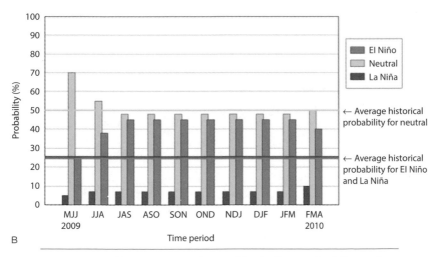

B

FIGURE 3.16 Presentation graphic used by the International Research Institute for Climate and Society to display the likelihood of an El Niño, La Niña, and neutral conditions in the (*A*) 2011–2012 ENSO season, and in the (*B*) for the 2009–2010 season. These are IRI probabilistic ENSO forecasts for the NINO3.4 region.

Source: International Research Institute for Climate and Society.

who are well. Sick children may be absent from school and cannot learn as well as their healthy classmates. Although this is true at any level of development, ill health is a classic ingredient of a poverty trap in poor countries. The poor are more likely to become sick due to lack of preventative health care services and unhygienic living conditions that promote disease. A person who has become ill is less able to earn or learn and is more likely to become even poorer, and this cycle perpetuates. This sort of poverty trap is a major impediment to advancing human welfare in many poor countries.

Figure 3.17 illustrates how closely related malaria prevalence and climate conditions are in Botswana. The figure shows SSTs

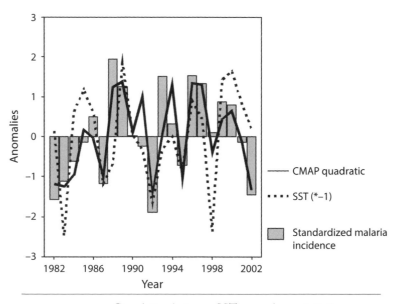

FIGURE 3.17 Correlation between SST anomaly, precipitation, and malaria prevalence in Botswana.

Source: Madeleine C. Thomson, Simon J. Mason, Thandie Phindela, and Stephen J. Connor, "Use of Rainfall and Sea Surface Temperature Monitoring for Malaria Early Warning in Botswana," *American Journal of Tropical Medicine and Hygiene* 73, no. 1 (2005): 214–221.

in the NINO3 area and rainfall (CPC-CMAP[8]) together with a measure of malaria incidence. The correlation between rainfall and SST is associated with the ENSO teleconnections.

The relationship to malaria incidence is equally striking. In Africa, malaria is commonly transmitted by the Anopheles mosquito, which carries *Plasmodium falciparum*, the most dangerous malaria parasite. The rate at which malaria is transmitted and developed is dependent on a number of climate-modulated environmental factors. Transmission occurs when a female mosquito bites an infected human and then survives long enough to bite another person once the parasite has developed. Both vector survival and parasite development are temperature dependent.

A temperature range from 18°C (65°F) to 32°C (90°F) is most suitable for development of the parasite, and humidity in excess of 60 percent is required for transmission through the mosquito host stage. Parasite development decreases significantly below 18°C, and above 32°C the survival of the mosquito is compromised. In addition, mosquitoes lay their eggs in standing water, and growth through the larval stage requires quite moist conditions (around 80 mm monthly rainfall at a minimum). Many effective efforts to eradicate malaria have involved draining swamps where mosquitoes breed. The strong dependence on temperature and standing water for parasite survival and growth leads to a moving geography of malaria risk (figure 3.18).

Conditions promoting malaria are common to the wet tropics of Africa. Regions where conditions are most conducive to malaria outbreak change with the seasons—darker areas in figure 3.16 indicate where malaria is stable. This north-south migration follows the ITCZ across the equator and back on a seasonal periodicity (see chapter 2).

Forecasting skills can now predict the development of environmental conditions suitable for aggressive malaria transmission

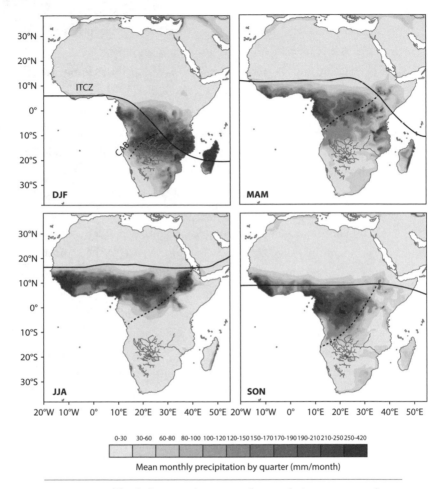

FIGURE 3.18 The dark area is the region where malaria is most prevalent and how it moves in a north-south sense with seasonal change in the relative location of the ITCZ.

Source: International Research Institute for Climate and Society.

several months in advance. ENSO forecasting enables us to anticipate warm moist conditions, and it provides an early warning for
health care systems that engage in malaria prevention. Although
suppression of the disease itself is generally not possible, sufficient
warning about anticipated outbreak conditions enables appropriate
medical interventions to be put in place to prevent large-scale outbreaks. *Malaria Early Warning Systems* (MEWS) bulletins issued
for those parts of Africa now incorporate seasonal forecasts of precipitation and temperature that might be conducive to an outbreak.

ENSO modulates a wide range of other processes important
for the advancement of human welfare in poorer countries. One
of the first recognized processes was the effect ENSO conditions
had on maize production in Zimbabwe (figure 3.19). This is one
of the most compelling correlations between a measured physical

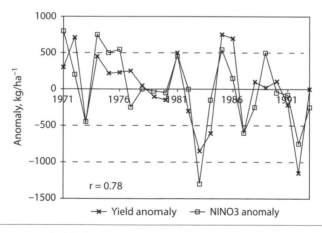

FIGURE 3.19 Graph shows a very high correlation between maize yield in
Zimbabwe and SSTs. The channel is through the effect of SSTs on rainfall
and hence crop yield. Forecasting Zimbabwean maize yield using eastern
equatorial Pacific sea surface temperature.

Source: Mark A. Cane, Gidon Ishel, and R. W. Buckland, "Forecasting Zimbabwean
Maise Yield Using Eastern Equatorial Pacific Sea Surface Temperature,"
Nature 370 (1994): 204–205.

parameter—SST values in the eastern Pacific (NINO3)—and crop yield (kilograms per hectare).

Strategies to anticipate and manage such swings in climate conditions include planting crop varieties most suited to the anticipated conditions, including seed varieties more resistant to drought or heavy rainfall. Crop insurance is also a valuable management tool when available.

OTHER SEASONAL AND INTERANNUAL CLIMATE SIGNALS

ENSO is one of a large number of quasi-periodic oscillations that create climate variability around the world. ENSO is distinguished by the significant level of prediction skill that has been achieved, allowing forecasts that have useful direct applications to development needs in poorer countries. The *North Atlantic Oscillation* (NAO) is another interannual variation, and as its name implies, it develops in the North Atlantic.

The NAO also has two phases. Figure 3.20 shows the atmospheric circulation patterns associated with each phase. Also similar to ENSO, NAO is described by an index associated with sea level pressure at two different locations. In this case, it is the pressure difference between Iceland and the Azores; hence it is a north-south difference rather than an east-west difference as was the case for ENSO. The same phenomenon of an oscillation between positive and negative states holds for both, but NAO peaks in April. In the positive phase of NAO, on the left in figure 3.20, the pressure difference is large. The pressure difference is considerably reduced in the negative phase, but it is still of the same sign (pressures are always higher in the Azores, and the index measures the magnitude of the difference).

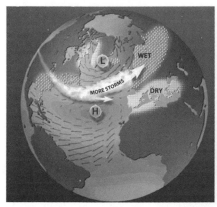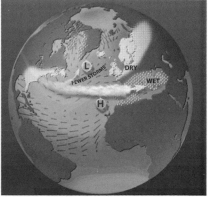

FIGURE 3.20 Distribution of sea level pressure systems in the
North Atlantic during positive (*left*) and negative (*right*) phases
of the North Atlantic Oscillation (NAO).

Source: Martin Visbeck/Lamont Doherty Earth Observatory,
http://www.ldeo.columbia.edu/res/pi/NAO/.

Unlike ENSO, NAO can remain in positive or negative phase for several years (figure 3.21). The last decade of the twentieth century was characterized by consistently positive NAO conditions. Earlier, in the 1960s and 1970s, the index had been persistently negative.

One of the principal effects of changing pressure gradients is a shift in the position of the polar jet stream (see figure 2.9), which affects the location of Atlantic storm tracks. In the positive phase, storm tracks tend to stay to the north, Europe experiences unusually wet (although not colder) conditions, and the Mediterranean region is unusually dry. In the negative phase, the reverse is the case, Europe is cold and dry, and the Mediterranean and North Africa are unusually wet.

Figure 3.22 shows historic stream flow averages for the Euphrates River, which has its headwaters in Turkey, at a time

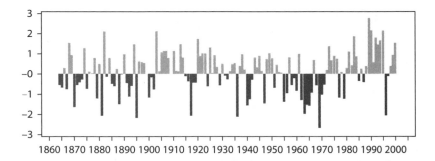

FIGURE 3.21 Time series (1864–2001) of the NAO Index showing how the value of the index may stay consistently positive or negative for many years.

Source: Jim Hurrell, Joint Institute for the Study of the Atmosphere and Ocean, http://research.jisao.washington.edu/data_sets/nao/.

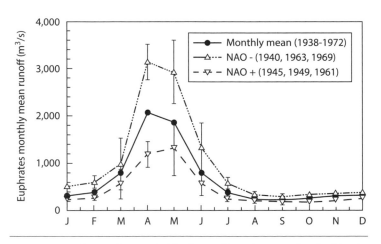

FIGURE 3.22 Stream flow in the Euphrates River (prior to dam building) correlates closely with the sign of the NAO Index.

Source: Heidi M. Cullen and Peter B. deMenocal, "North Atlantic Influence on Tigris-Euohrates," International Journal of Climatology 20, no. 8 (2000): 853–863.

before most of the dam building on the river took place. It shows significant differences in flow associated with the two phases of NAO. The Euphrates is a critical source of water for irrigation, especially in Syria and Iraq, and variations in flow of this magnitude bring considerable risk to crop production and enhance social stress.

In North America, the principal effect of NAO is to modulate conditions in the southeast and eastern sections of the country, where it has a considerably stronger influence on weather than does ENSO. In addition, evidence suggests that tracks of Atlantic hurricanes are influenced as NAO alters conditions where these storms are generated, and influences where they travel. The second factor may be the most important. Hurricane tracks in the Atlantic basin often trend westward initially, then hook northward under the influence of the Coriolis effect. However, the track is strongly influenced by a high pressure system that develops in the Atlantic during hurricane season. The system is known as the Bermuda-Azores high, and it causes the tracks of hurricanes, especially those that develop off Cape Verde in the latter part of the hurricane season, to stay to the south for a greater distance than they otherwise would. In effect, the hurricanes skirt around the high pressure system.

One consequence when the Bermuda-Azores high is very strong is that hurricanes may experience the northward hook farther west; therefore, they are more likely to make landfall. When the Bermuda-Azores high is strongest, as it is in NAO's positive phase, more hurricanes will intersect the Caribbean and the U.S. east coast. When NAO is negative, and the Bermuda-Azores high is weaker, hurricanes turn north in the Atlantic Ocean and are more likely to stay at sea. The effect of NAO on Atlantic hurricanes does not involve changes in wind shear, as is the case with ENSO conditions that inhibit the formation of hurricanes, but it steers hurricane tracks after their formation.

Finally, in describing NAO, no oceanographic phenomena are currently known to be involved, such as the change in thermocline depth essential to understanding ENSO. NAO is essentially an atmospheric oscillation, and it does not have an associated expression in ocean processes. That is one factor that limits its predictability.

The Pacific Ocean has a somewhat similar feature, the *Pacific Decadal Oscillation* (PDO), and it develops in a north-south direction as well. It, too, is measured by a similarly derived index, and it is predominantly an atmospheric signal. One of the most important effects of the PDO is that it appears to modulate the ENSO phase (figure 3.23). When the PDO is positive, El Niño conditions are more prevalent and may remain so for a decade or more, similar to the NAO. Also like the NAO, no prediction skill has been achieved at present for the PDO.

As a final note, although these oscillations do not develop on a fully global scale, they have a considerable geographic extent. Global effects of these oscillations may enhance or mask effects of GHG-induced warming. In the first decade of the twenty-first century, global warming appeared to more or less stall, and it is

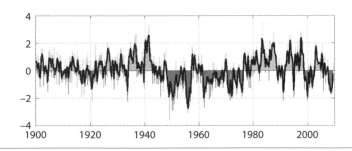

FIGURE 3.23 Monthly values for the PDO index, 1900–2009. The PDO has long periods of consistently positive or negative values that may influence the phase of the ENSO.

Source: Joint Institute for the Study of the Atmosphere and Ocean.

considered likely that one or a combination of these large-scale oscillations was responsible. A likely explanation comes from the fact that La Niña conditions were more prevalent in this period than is typical in a very long-term sense. In addition, the PDO was in a persistent negative phase during this period, which may have added to the stall-out in warming. The stall has now ended, however, and temperatures are steadily rising again.

4

CLIMATE IN THE FUTURE

As described in the discussion of ENSO forecasting in chapter 3, climate predictions are achieved using computer models. These models represent the behavior of natural systems, including projected future behavior, using highly simplified mathematical descriptions of natural processes. For climate prediction, the model is computer code, sometimes referred to as a numerical simulation. Almost all of the familiar weather forecasts provided by media are made using computer models. One of the great advantages a computer has is a vast and unfailing memory.

Deep Blue, IBM's chess-playing computer, was able to beat Garry Kasparov because it stored thousands of games in memory, compared the current game to its database, and tried millions of next moves to determine which would provide the greatest advantage. The computer simulated new games that might follow a suite of next moves, in effect producing model chess games. Kasparov simply did not have the memory, or speed of thought, to simulate millions of possible moves, even though his IQ has been measured at 190.

CLIMATE MODELS

Global climate models are different from weather forecasts in very important ways. One difference is that the past provides only a limited guide to the future into which climate predictions are made. The models need to project into a future in which greenhouse gas (GHG) concentrations are much greater than they have been in the past few thousand years, so they enter a territory for which there is little precedent. To state that difference more clearly, weather forecasting is an initial value problem—present conditions are constantly updated. Climate predictions are a boundary value problem—after initialization, the model outcome is governed by changes in prescribed conditions such as GHG concentration, and the model outcome becomes largely independent of the starting position.

A comprehensive climate model includes a large and complex set of mathematical expressions (comprising sets of simultaneous differential equations in continuous form solved by standard numerical methods). These expressions nevertheless describe idealizations. For instance, the Navier-Stokes equations (four simultaneous differential equations) accurately describe how convection operates and gives rise to Hadley cells and so forth, but it is not possible to account for the development of clouds, precipitation, or chemical reactions in the atmosphere using these equations.

In fact, the global climate model is not solved by the full form of the Navier-Stokes equations but by a reduced form known as the "primitive equations." These equations accurately describe an idealized, unrealistic situation in which no clouds form and it never rains. In effect, they accurately describe a low-level approximation of the world, just as the simple mathematics we used to explain why Earth is about 15°C on average accurately describes a very simplified Earth. The basic equations can be modified to include many more processes that affect climate, but they are generally introduced as add-ons and can be very cumbersome. Technically this is known as *parametrization*; that is, some processes are introduced but not calculated by the model itself. Adding features means that the depiction of Earth in the model becomes a better approximation of the real Earth, but it remains an approximation.

Global climate models have become more comprehensive over time, incorporating more and more aspects of Earth's systems that are important in climate processes. The earliest models from the 1970s didn't even include the oceans. When oceans were first added, they effectively had no depth and were referred to as swamp oceans. It is only in the most recent models that such features as ocean circulation, response of vegetation to changing climate, clouds, and chemical reactions in the atmosphere have been incorporated.

BOX 4.1 THE INTERGOVERNMENTAL PANEL ON CLIMATE CHANGE (IPCC)

The IPCC, founded in 1988 by the World Meteorological Organization and the United Nations Environment Program, is an international body assessing the state of science related to climate change. The panel regularly assesses developments, future risks, and improvement efforts in the realm of climate change research. It currently has 195 member countries.

Its purpose is to promote a consensus for all countries involved on the implications of climate change, so governments may develop informed policy for climate action:

The IPCC embodies a unique opportunity to provide rigorous and balanced scientific information to decision-makers because of its scientific and intergovernmental nature.

IPCC work is shared between three working groups:

Working Group I: assesses the physical scientific aspects of the
climate system and climate change.
Working Group II: assesses the vulnerability of socioeconomic and
natural systems and options for adaption to negative impacts.
Working Group III: assesses options for mitigating climate change
through limiting or preventing greenhouse gas emissions and
enhancing activities to remove them from the atmosphere.

The IPCC has thus far issued five "Assessment Reports" and
the sixth is underway. All reports are available at www.ipcc.ch
/report/ar5/.

Even so, these models remain approximate and incomplete depic-
tions of the planet.

Along with the equations that describe the mathematical
physics of climate in computer code, Earth itself has to have a
representation in the computer. To do this, Earth is divided into
grid cells (figure 4.1). The finer the grid cell, the more realisti-
cally Earth can be represented. The figure shows the improvement in resolution obtained in successive IPCC models. This
is something similar to pixel density of a computer screen or
television, although no climate model can be thought of as having high definition. No model can derive any aspect of climate
on a spatial scale smaller than the grid size. The grids shown
represent Earth's surface as an example, but the atmosphere and
oceans must also be represented in gridded form.

The grid size in current models is much finer than earlier generations of models, but the size is still too large to adequately

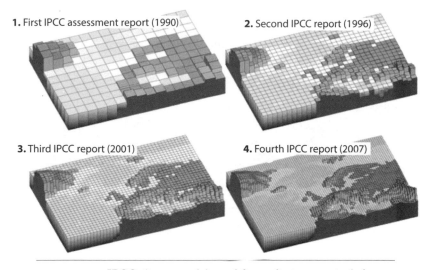

1. First IPCC assessment report (1990)

2. Second IPCC report (1996)

3. Third IPCC report (2001)

4. Fourth IPCC report (2007)

FIGURE 4.1 IPCC climate models used for prediction must include a representation of the Earth's topography and numerous other parameters. These are input in grid form in which parameter values are constant in each grid cell. Over time the representation has become more detailed. In the 1990 report the shape of Europe is barely recognizable.

Source: Intergovernmental Panel on Climate Change.

represent a number of features important in the climate system. Cloud formations, for instance, are often much smaller than the 110 × 110 km grid size of current models. Rivers are also much smaller than this grid. Vegetation type varies considerably at scales much smaller that the finest model grids. The effects of these, and a variety of other processes, are approximated and introduced as an average over the grid cell.

Why not use even smaller grid sizes? The reason is purely practical. A computer calculation is made using equations that describe a simplified system for each cell. More cells require more calculations, and each one takes a finite, though tiny, amount of time. The more complete the model, the more terms and parameters are required in the equations. More cells, combined with more complete representations of the physics and chemistry of climate processes, requires more time to execute a model simulation. The most important reasons climate models have become higher in resolution, and more complete, are that computers are constantly getting faster and memory is getting larger.

For example, if a model makes a simulation of the future climate every ten years for one hundred years, that calculation consumes 10 times less computer time than making a simulation every year for one hundred years. Current models run at time resolutions as fine as a few minutes, so the number of calculations needed to project to 2100, for example, is enormous.

All the major climate research institutions—NASA, NOAA, and NCAR in the United States and the Hadley Center in the United Kingdom—have very large computer facilities dedicated to making predictions. Significant efforts have been made in model intercomparison, running suites of models under specified conditions and diagnosing differences in output, so models can be more consistent in their performance. Nevertheless, climate model outputs are destined to be approximations that predict changes in an idealized world, and idealizations will differ from one model to another.

Surely there is only one correct climate model. That would be ideal, but no agreement exists today on that model. This is somewhat akin to evaluating search engines. We are all familiar with searching on keywords. But if you use the same keyword with different search engines, you will get different answers, at least after a page or two of searching. Why? The search algorithm associated with each search tool is different, maybe not hugely different but different enough to give varied answers. And those answers become more different as the search progresses. The first ten results of a keyword search by different tools may be quite similar, but after the fiftieth search the results will look quite different—not unlike the way climate models diverge more as they predict farther into the future. Is there one correct way to search? Probably not.

HOW PREDICTIONS ARE MADE

A typical model scheme can be depicted as a cascade, or sequence, of steps as shown in figure 4.2. The model itself is shown shaded.

Model Inputs: Emission Scenarios

The heart of a climate model is a calculation that takes prescribed changes in the concentration of numerous constituents of the

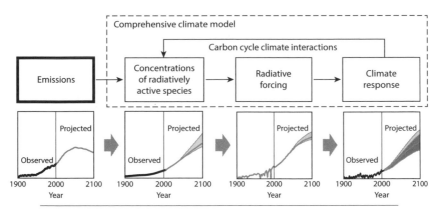

FIGURE 4.2 A climate model makes calculations of Earth's future temperature using as input the projected change in emissions of GHGs. The model calculation proceeds in several steps explained in the text.

Source: "Global Climate Projections," chap. 10 in Intergovernmental Panel on Climate Change, *Climate Change 2007: The Physical Science Basis*, contribution of Working Group I to the *Fourth Assessment Report of the Intergovernmetnal Panel on Climate Change*, ed. Susan Solomon, Dahe Qin, Martin Manning, Zhenlin Chen, Melinda Marquis, Kristen B. Averyt, Melinda M. B. Tignor, and Henry LeRoy Miller Jr. (Cambridge: Cambridge University Press, 2007).

atmosphere and determines the temperature change that would result, together with other derived features of climate such as precipitation. The model requires concentrations of these atmospheric components both at the present time and into the future. Models do not calculate the future concentrations of anthropogenic GHGs but require them as an input. The most recent models, including biochemical and atmospheric chemistry, do calculate changes in natural GHG concentrations, but the anthropogenic component must be specified as a boundary condition.

Climate model inputs are described in terms of *emissions scenarios* that describe how GHG emissions might evolve in the future. They rely on different projections of factors such as population growth, the adoption of new cleaner technologies for energy production, changes in development status around the world and associated changes in consumption, and energy use. The IPCC uses a suite of scenarios and publishes a special report describing how they arrived at each scenario and listing the types of components that go into each scenario. Each scenario is given an acronym and represents a different conception of what the future might be like. Box 4.2 outlines the types of components that go into each scenario. Figure 4.3 illustrates the importance of scenario choice.

Some scenarios are quite pessimistic about the future, suggesting that emission rates will remain high throughout the twenty-first century and population will continue to increase. Other scenarios are more optimistic, imagining a peak in emissions at midcentury and large-scale adoption of new cleaner energy technologies. For instance, the second panel in figure 4.3 predicts a 6°C temperature change, whereas the fifth panel model predicts about a 2°C temperature change. The difference in the calculations reflects different inputs to the model for calculation. Critical to the construction of these global scenarios are assumptions about how the poorer economies will develop and the emissions consequences that might arise from that.

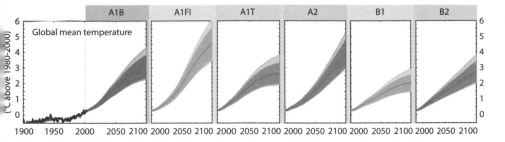

FIGURE 4.3 Temperature predictions made using different scenarios and the same suite of models.

Source: "Global Climate Projections," chap. 10 in Intergovernmental Panel on Climate Change, *Climate Change 2007: The Physical Science Basis*, contribution of Working Group I to the *Fourth Assessment Report of the Intergovernmetnal Panel on Climate Change*, ed. Susan Solomon, Dahe Qin, Martin Manning, Zhenlin Chen, Melinda Marquis, Kristen B. Averyt, Melinda M. B. Tignor, and Henry LeRoy Miller Jr. (Cambridge: Cambridge University Press, 2007).

BOX 4.2 EMISSIONS SCENARIOS FROM THE SPECIAL REPORT ON EMISSIONS SCENARIOS (SRES) AR4

A1. The A1 storyline and scenario family describes a future world of very rapid economic growth, global population that peaks in mid-century and declines thereafter, and the rapid introduction of new and more efficient technologies. Major underlying themes are convergence among regions, capacity building and increased cultural and social interactions, with a substantial reduction in regional differences in per capita income. The A1 scenario family develops into three groups that describe alternative directions of technological change in the energy system. The three A1 groups are distinguished by their technological emphasis: fossil intensive (A1FI), non-fossil-energy sources (A1T), or a balance across all sources (A1B) (where balanced is defined as not relying too heavily on one particular energy source, on the assumption that similar improvement rates apply to all energy supply and end-use technologies).

A2. The A2 storyline and scenario family describes a very heterogeneous world. The underlying theme is self-reliance and preservation of local identities. Fertility patterns across regions converge slowly, which results in continuously increasing population. Economic development is primarily regionally oriented, and per capita economic growth and technological change are more fragmented and slower than other storylines.

B1. The B1 storyline and scenario family describes a convergent world with the same global population, peaking in midcentury and declining thereafter, as in the A1 storyline, but with rapid change in economic structures toward a service and information economy, with reductions in material intensity and the introduction of clean and resource-efficient technologies. The emphasis is on global solutions to economic, social, and environmental sustainability, including improved equity, but without additional climate initiatives.

B2. The B2 storyline and scenario family describes a world in which the emphasis is on local solutions to economic, social, and environmental sustainability. It is a world with a continuously increasing global population at a rate lower than A2, intermediate levels of economic development, and less rapid and more diverse technological change than in the A1 and B1 storylines. Although this scenario is also oriented toward environmental protection and social equity, it focuses on local and regional levels.

Source: Intergovernmental Panel on Climate Change

In highly developed countries like the United States, Canada, Australia, and the EU countries, individuals contribute far more carbon dioxide to the atmosphere than do individuals in poor countries. In large part, this is due to emissions from the power sector and manufacturing, which are typically less advanced in poor countries. So-called mobile power, meaning fuels in transportation, are an important source of GHGs, and that can be important in poorer countries.

An important issue that arises for calculations of future emissions is that poor countries have strong and legitimate aspirations for welfare improvement, and in the past that has depended on development of energy resources based on fossil fuels. If poor countries mimic today's wealthy countries in their development strategies, and developed countries do not significantly reduce emissions from fossil fuels, the most pessimistic of all emissions scenarios will come about with GHG concentrations far more than double their concentration today.

One critical factor in constructing emissions scenarios is population growth. Even if some people are not emitting a great deal, an increasing population will correspond with increasing emissions. Almost a threefold difference in population projections to 2100 is shown between the AR4 scenarios, and this has a profound effect on emissions. A large population in itself does not necessarily imply vastly greater emissions. High fertility rates are typically associated with poor countries today where per capita emissions are very low. More important is the growth in population of people whose lifestyles are associated with high levels of fossil fuel use for energy and transportation.

Figure 4.4 shows AR4 emissions scenarios for carbon dioxide. The range in carbon dioxide emissions across these scenarios is even greater than the range of population estimates, in percentage terms.

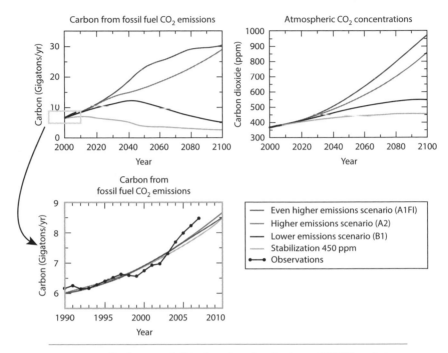

FIGURE 4.4 In the upper left is the suite of projections of GHG emissions used as input to climate models. In the upper right, emissions have been converted to concentrations. Emissions for the four scenarios in the recent past are shown in the lower left.

Source: Based off of U.S. Global Change Research Program, *Global Climate Change Impacts in the United States 2009 Report*, https://nca2009.globalchange.gov/index.html.

Emissions are usually described in gigatons per year (Gт/yr); a gigaton is a billion tons. Emissions are a flow rate, like the strength of water running from an open faucet, and therefore the unit is expressed as an amount per unit of time, usually one year. Note that in the figure the axis is in gigatons of total emissions. It is important to note that even the optimistic scenarios suggest that emissions will begin to decline in midcentury and concentrations will continue to increase, if more slowly.

From Emissions to Concentrations

There is another popular misunderstanding concerning emissions reduction targets. If emissions rates were reduced to 1990 levels, that would not reduce global average temperature to 1990 values. It would only reduce the rate of temperature rise to 1990 rates. The only way temperature will stabilize is by reducing emissions to very near zero. The atmosphere responds to the total concentration of GHGs, and that is derived from the emissions as a first step (see figure 4.2).

As a first step, the model calculates concentrations based on emissions scenarios, one for each scenario. This calculation involves some uncertainty because the exact proportion of each GHG into various sinks—atmosphere, biosphere, and ocean—is not precisely known.

From Concentrations to Radiative Forcing

With the concentrations calculated from differing emissions scenarios, the next step is to estimate the effect this will have on near-surface air temperature. This requires the calculation of *radiative forcing*. How much does a given change in the concentration of GHGs change the critical balance of incoming and outgoing radiation? The calculation is usually made at the level of the troposphere. If net forcing is positive, Earth's surface temperature must rise to move the balance back to equilibrium. If it is negative, Earth must cool.

Figure 4.5 shows the best estimates available of the forcing effects of various components of the atmosphere relative to pre-industrial values.

Here another uncertainty enters into model calculations. Attached to each bar that represents the magnitude of forcing

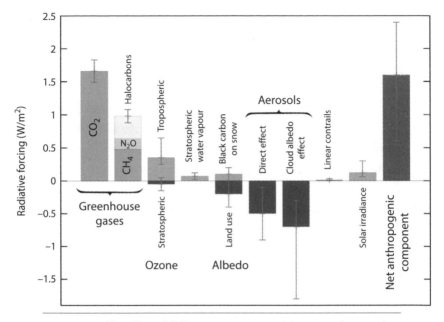

FIGURE 4.5 The effect of different components of the atmosphere on the balance of incoming and outgoing radiation—radiative forcing. GHGs all have positive forcing so increasing their concentration causes an imbalance that leads to warming. Aerosols have negative forcing effects. The net or aggregate forcing is about 1.5 watts per square meter in a positive sense but with a high range of uncertainty.

Source: Courtesy of Leland McInnes/Wikimedia Commons.

of a particular atmospheric component is a thin vertical line with a short horizontal bar at either end. This expresses the uncertainty in the forcing estimate—sometimes called whiskers. The forcing effect of some components, such as CO_2, is quite accurately known (the whiskers are short) and has been measured in lab experiments, but the effect of others (ozone, for instance) is much less certain (long whiskers).

Chief among those forcing components about which there is large uncertainty are *aerosols*, which often result from the polluting effect of incomplete fossil fuel combustion at low temperature, as happens, for instance, in diesel engines. They are known to have negative forcing, meaning that additions of these components will lead to cooling rather than warming. Aerosols are not GHGs; they are solid particles, and their effect is to shield Earth from incoming radiation. GHGs are usually defined as gases that give rise to absorption of infrared radiation emitted by Earth. The size of the cooling effect of atmospheric aerosols could be quite large, possibly as large as the warming effect of CO_2. The large uncertainty associated with aerosols comes from the many ways in which they interact with incoming radiation (seven ways in all). Aerosols also have direct and indirect effects on cloud formation and properties, each of which has a different magnitude although all are negative. The effect of aerosols might, for instance, enhance the lifetime of some clouds. Note that the aerosol effect is not a feedback; it is the direct result of fossil fuel combustion.

All forcings have been added together and a "net anthropogenic forcing" is determined (most right-hand column in figure 4.5). The net effect is positive, so overall warming is implied, but with a large range of possible values—from as little as 0.5 to as much as almost 2.5 w/m^2. The larger figure would imply that aerosols have a relatively small effect and that GHGs dominate radiative forcing. The smaller figure would imply that negative forcings resulting from aerosols are almost in balance with positive forcings from GHGs.

The authors of the IPCC fifth assessment report acknowledged the importance of emphasizing radiative forcing and chose to represent scenarios in terms of representative concentration pathways (RCP; figure 4.6). The concentration of GHGs in the atmosphere is the primary concern, and targets should be thought of in those terms rather than emissions. A reduction in emissions

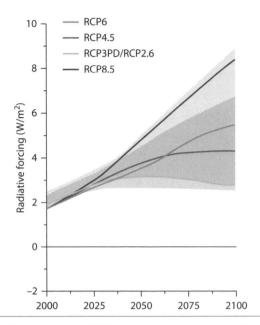

FIGURE 4.6 The most recent IPCC report, Assessment 5, uses Representative Concentration Pathways (RCPs).

Source: Detlef P. van Vuuren, Jae Edmonds, Mikiko Kainuma, et al., "The Representative Concentration Pathways: An Overview," *Climatic Change* 109 (2011): 24. DOI: 10.1007/s10584-011-0148-z.

does not imply a reduction in concentration—something that is often missed in policy discussions.

RCPs essentially combine the first step of a climate model in which emissions are used to calculate concentrations. The vertical axis in figure 4.6 is now scaled in units of w/m^2 and is the radiative forcing, not a concentration. In RCP8.5, for instance, the forcing at the end of the century is 8.5 w/m^2. That is, the radiative balance would be altered by 8.5 w/m^2 and would require Earth to warm considerably to restore the balance. Another advantage to framing the discussion using RCPs as a metric is that temperature is linearly related to the cumulative total of anthropogenic emissions.

Comparing the effect of different GHG components of the atmosphere is frequently done using the *global warming potential* (GWP), which compares the effect of these gases over specified periods of time. Atmospheric components have quite variable lifetimes (sometimes called *average residence times*), and they remain in the atmosphere for different periods of time. GWP describes the ability of a specified gas to absorb and reradiate heat compared to carbon dioxide over a specified period of time, from 20 to 500 years. The GWP of carbon dioxide is always set at "1," and other greenhouse gases are compared to carbon dioxide for the same time frame. For example, methane has a GWP of 56 integrated over 20 years, 21 over 100 years, and 6.5 over 500 years. The most potent of all GHGs is sulfur hexafluoride (SF6), which has an atmospheric lifetime of 3,200 years; the 20, 100, and 500 year GWP values are 16,300, 23,900 and 34,900, respectively. Fortunately, the concentration of SF6 in the atmosphere is very small; there are no natural sources, and human activity is no longer increasing its input to the atmosphere. Although water vapor is a strong GHG, it does not have a calculable GWP because it does not decay in the atmosphere.

Calculating the Temperature Field

The final step is to calculate temperature from estimated forcing. In principle, this should be straightforward, with the caveat that there are considerable associated uncertainties in how to treat various feedback effects. Of the many feedback effects in the climate system that contribute to the uncertain outcome of climate models, one of the most studied and potentially the most important is the effect of clouds—*cloud radiative feedback* (figure 4.7).

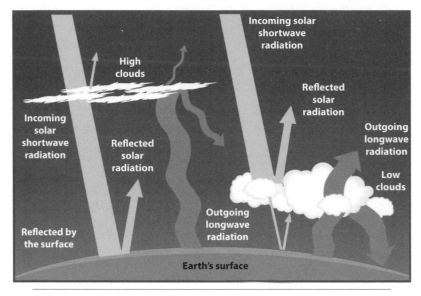

FIGURE 4.7 Cloud radiative feedback. Low clouds are generally very reflection and hence have a cooling effect on the Earth's surface while highs clouds have the opposite effect.

Source: NASA/Visible Earth, "Cloud Effects on Earth's Radiation," https://visibleearth.nasa.gov/view.php?id=54219.

Clouds are composed of water vapor, water, and ice (H_2O in gas, liquid, and solid form). Water vapor condenses on cloud concentrating nuclei to form ice and water. Clouds have a major effect on maintaining the temperature of the planet at 15°C in two ways. Water vapor is a GHG, so it has a positive feedback effect on near-surface temperature. But clouds also may have high albedo and reflect a considerable amount of incoming short-wavelength radiation (see chapter 2), so clouds can have positive or negative forcing. Which effect dominates depends on the cloud type and its altitude (see figure 1.13). In general, low cumulus clouds are quite strongly reflective (high albedo), and their dominant effect in the climate system is negative feedback cooling. High clouds are the opposite. They are very thin with low albedo—the Sun may be visible through high clouds—and their dominant effect is positive feedback warming.

As the world warms, more clouds will form because more ocean water will evaporate. Setting aside clouds for the moment, adding water vapor to the atmosphere itself will amplify any temperature increase because moister air holds more heat—a positive feedback. With more clouds, there will be more of the two types of cloud feedback effects. If the cooling effect wins over warming, then clouds will have an overall countering effect to warming, pushing temperatures back down as cloudiness increases. If high clouds dominate, they will have a net warming effect, leading to even more warming. The first is a negative feedback effect and the second is a positive feedback effect, and both are close to instantaneous. Of twenty commonly used models, fourteen have negative aggregate cloud feedback effects and six have positive cloud radiative feedback effects. The sign and strength of cloud radiative feedback is the source of the greatest uncertainty in climate change model predictions today.

The reflectivity of low-level clouds has inspired one of many propositions to intervene in Earth's radiative balance (with particles introduced into the atmosphere via balloon in one suggestion) to shield Earth from solar radiation. This is known as *solar radiation management* (SRM). Many other ideas have been proposed, including stripping carbon dioxide directly from the atmosphere. Collectively these ideas are described as *geoengineering*, a topic too large to cover in this book.

Different models give different results, even when run with the same input scenario (figure 4.8). These model runs begin by "predicting" the past. The model simulations start in 1850 and display the slow variations that characterize the preindustrial period. They also capture the rise in temperature in the immediate postindustrial period, but they diverge soon after that time.

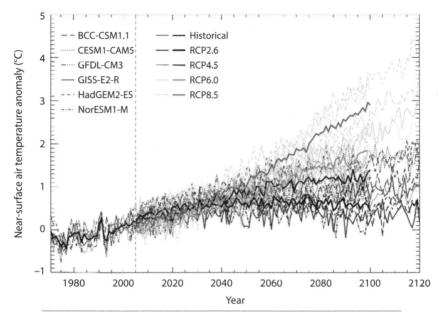

FIGURE 4.8 Different computer models provide different temperature predictions even given the same parameters as input. They diverge more with time into the future. Those that predict in the higher range are described as "sensitive" and in the lower range "insensitive" to GHG forcing.

Source: New Zealand National Institute of Water and Atmospheric Research, "Climate Change Scenarios for New Zealand," https://www.niwa.co.nz/our-science /climate/information-and-resources/clivar/scenarios.

Because both scenario choice and model play important roles in climate predictions, it can be useful to illustrate both in the same diagram. Figure 4.9 shows the results of six models using two different scenarios: the most optimistic and the most pessimistic of the AR4 suite of scenarios. There is a sixfold difference in the derived year 2100 temperature prediction, from less than 2°C to almost 6°C.

Temperature change as expressed in figure 4.9 is the average air temperature near the surface of Earth, but the change will not be

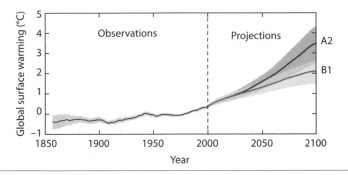

FIGURE 4.9 A typical way in which temperature projections are illustrated combines a suite of scenarios with a suite of models to give an aggregate estimate of temperature change. Here there are two scenarios shown, A2 and B1 and the spread of projections is that arising from different models.

Source: Reto Knutti, "Should We Believe Model Predictions of Future Climate Change?," *Philosophical Transactions of the Royal Society A*, September 25, 2008, https://doi.org/10.1098/rsta.2008.0169.

uniform throughout Earth. All models show that the high northern latitudes will experience much greater changes in temperature than low to middle latitudes because they are currently ice covered; the ice albedo feedback described previously enhances the warming effect. In the more pessimistic scenario/model combinations, temperatures rise by almost 10°C in the Arctic. There is ample evidence that this is indeed happening.

Figure 4.10 illustrates a different representation that includes a rendering of the changes across the globe for two different RCPs: one very optimistic (2.6), one quite pessimistic (8.5).

Figure 4.10 is a common way to illustrate future temperature projections. Rather than select a midrange scenario, a suite of scenarios is chosen that express the range of plausible outcomes. In this way, the projections can be thought of as best-case and worst-case scenarios. What this type of representation cannot do

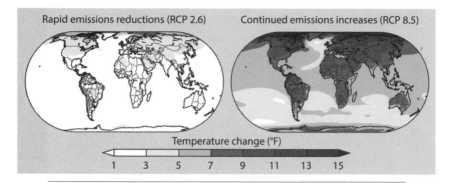

FIGURE 4.10 Projected change in average annual temperature for two RCP scenarios.

Source: Jerry M. Melillo, Terese Richmond, and Gary W. Yohe, eds., *Climate Change Impacts in the United States: The Third National Climate Assessment* (Washington, D.C.: U.S. Government Printing Office, 2014), doi:10.7930/J0Z31WJ2.

is provide any sense of uncertainty in the projections that could be assessed from figures 4.8 and 4.9.

POSSIBLE FUTURES

In summary, two sources of uncertainty give rise to predictions that range from 1°C to more than 6°C by 2100. The first uncertainty is the very large range in emissions scenarios (RCPs), and the second is inherent uncertainties in model outputs. Scenario uncertainty is sometimes referred to as boundary value uncertainty because the emissions described by a scenario provide the boundary values for the model calculation. The two uncertainties combine additively. The great uncertainty in human behavior implicit in the scenarios/RCPs introduces by far the greater uncertainty.

If it were somehow possible to decide on a "correct" scenario, the range of model-predicted temperatures would be reduced substantially. For instance, if it were decided that RCP 2.5 was "correct"—or the one used for planning purposes—the range of predictions for 2100 would be reduced to 2°C to 3.5°C, less than half of the total range found across all scenarios and models. Similarly, if it could be decided that one model is best, that too would shrink overall uncertainties.

One simple way to think about the outcomes of model calculations is in terms of *climate sensitivity*. We can write this:

$$\Delta T = \lambda \Delta F$$

ΔT is the change in temperature that would result from a change of ΔF in the forcing. The factor λ is described as climate sensitivity (this is not the same λ as that used for wavelength in describing electromagnetic waves).

If λ is a large number, relatively small changes to the properties of the atmosphere will have large effects on surface air temperature—the climate is sensitive to small changes in GHGs. If the value is small, the atmosphere can change a lot with little effect on surface temperatures—the climate is relatively insensitive to GHGs. The value of λ implicit in model calculations is therefore extremely important in making projections of climate into the future.

How is the value of λ obtained? One way is to examine the ancient record. The best archive of ancient climate is found in ice cores. They record both temperature in the oxygen isotope ratios of the ice crystals and trap tiny bubbles of the ancient atmosphere, which can be analyzed to describe the composition of the atmosphere at the time the ice formed. The two upper graphs in figure 4.11, obtained from analysis of ice cores taken in Antarctica, show the concentration of carbon dioxide and methane, two very important GHGs. The lower graph is the temperature derived from oxygen isotope analysis. The front cover of this

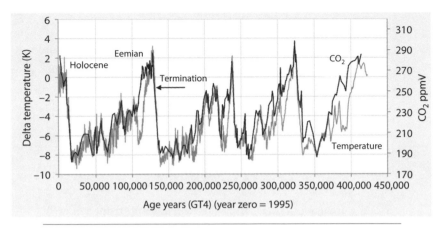

FIGURE 4.11 A comparison of temperature, carbon dioxide (CO_2) concentration obtained from ice core data in Antarctica.

Source: NOAA, http://www.ncdc.noaa.gov/paleo/icecore/antarctica/vostok/vostok.html.

book displays one such sample of ice, and you can see the bubbles of the ancient atmosphere trapped inside the dark spots in the ice.

The high degree of correlation between these curves is striking, and this can be used to estimate climate sensitivity, assuming that the direction of causation is from GHG concentration to temperature change rather than the other way around. Using this approach, sensitivity can be estimated to be $3/4°C \pm 1/4°C$ per w/m². The forcing (ΔF) is expressed in w/m², which is the same unit as energy flux discussed in chapter 2. The energy flux coming into the top of the atmosphere is 350 w/m² averaged over the whole Earth, and climate forcing is described as the change relative to that background level.

Sensitivity also can be estimated by running models and matching their output to records of past climate obtained from proxy data such as tree rings. One way to express sensitivity is to ask what temperature change at the surface would be expected for an instantaneous doubling in carbon dioxide concentration.

The range of sensitivities estimated from various approaches is 1.5°C to 4.5°C. A commonly used value is 3°C, meaning that were the carbon dioxide content of the atmosphere to double, the expected change in temperature after equilibrium has been reached would be 3°C.

There are a few important caveats to consider. One is that the sensitivity equation assumes a linear relationship between forcing and temperature, which cannot be substantiated. Second, much of what will determine the value of the sensitivity parameter λ are feedback effects that operate on very different time scales: some very short, others much longer. It is common to consider two sensitivities, one that is the immediate response to a doubling of CO_2, referred to as instantaneous climate sensitivity, which is the immediate change that would come about from a sudden change in CO_2. The second is equilibrium climate sensitivity, which is the relationship after all feedback effects come into play and the climate has reached equilibrium. Estimates of sensitivity from ice core records provides this measure.

Another measure of sensitivity is the *transient climate response to cumulative carbon emissions*, which is the mean surface temperature change that would come about per 1,000 *GTC*. It allows the discussion to be framed in terms of total accumulated carbon in the atmosphere, and it has the advantage of a linear relationship to near-surface temperature.

One way to summarize the interacting roles of climate sensitivity and scenario outlook is shown in figure 4.12. Two extreme outcomes are considered. In the lower left is the most desirable outcome. It uses an optimistic emission scenario in which GHGs are steadily reduced and economic growth is equitable. It involves modest population growth combined with an insensitive climate. In the upper right is a pessimistic scenario in which emissions remain high combined with a very sensitive climate system.

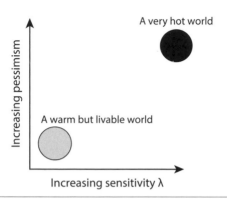

FIGURE 4.12 The interacting roles of climate sensitivity and scenario outlook.

Any location in the space of this sketch is possible, depending on the choice of model and scenario. More important, it is not reasonable to assume that a midway point between the two extremes would be the most likely future.

Past experience does not provide clear guidance on how to predict the future using models of this sort. The past record is, of course, useful in providing guidance on factors that lead to different climate conditions, but this information does not easily lead to improved model prediction strategies. Examination of past climate is very important, however, in estimating equilibrium sensitivity. Statistical predictions of ENSO (and daily weather), for instance, benefit from examination of a time series of many previous El Niño and La Niña events, and weather system models are improved and updated with every new occurrence. One of the defining characteristics of the global climate prediction problem is that Earth is entering a state in which GHG concentrations are higher than any levels experienced for millions of years, so past experience cannot be used directly to tune climate models.

In addition, there is no target time for prediction. Unlike El Niño, which always peaks in December, defining the "peak" of climate change would be arbitrary. Many predictions are made for the year 2100, but that does not mean climate change will have peaked at that time.

As a final comment on uncertainties, consider figure 4.13, which shows a compilation of estimates of equilibrium climate sensitivity. The horizontal axis is the expected temperature, and the vertical axis is the probability of that expectation. The compilation is derived from the analysis of many papers in the scientific literature, each of which makes an assessment of sensitivity. What is of great importance is the overall non-Gaussian shape of the distribution (box 4.3).

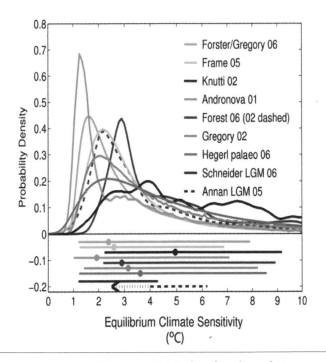

FIGURE 4.13 Climate sensitivity based on the analysis of twenty literature studies.

Source: NASA Earth Observatory.

BOX 4.3 SKEWED GAUSSIAN DISTRIBUTION

A skewed Gaussian distribution is compared here to an unskewed Gaussian, or normal, distribution. The normal distribution is symmetric about its peak value, as shown in the example to the left. The skewed distribution is asymmetric, with one side much more extended than the other. In the example shown, the distribution would be said to be right-skewed because the distribution appears to be extended to the right side. This is sometimes called a fat-tailed distribution.

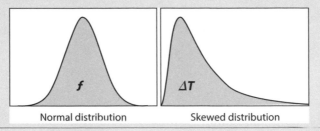

| Normal distribution | Skewed distribution |

BOX FIGURE 4.3.1 Gaussian distributions.

The reason the distribution is fat tailed comes entirely from the effects of feedbacks, as explained in this illustration:

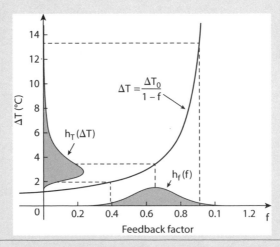

BOX FIGURE 4.3.2 Gaussian uncertainty and aggregate feedback.

Source: Gerard H. Roe and Marcia B. Baker, "Why Is Climate Sensitivity So Unpredictable?," *Science* 318, no. 5850 (2007): 629–632.

Shown on the horizontal axis is the aggregate system feedback, f, assumed Gaussian in its distribution and with a positive mean as in the upper panel. The curve in the middle of the graph is the change in temperature from a starting state T_o for a given feedback, f. The vertical axis then maps the Gaussian uncertainty in feedback factors into a right-skewed, or fat-tailed, probability distribution of temperature outcomes. A fat-tailed distribution will always result if the net feedback is positive and Gaussian. This mapping applies to any system with positive aggregate feedbacks and is not unique to the climate system.

Gaussian distributions, or so-called bell curves, are perfectly symmetrical and describe random processes. A "fat-tailed" distribution[1] is asymmetric—one side of the distribution trails off slowly, and the other side drops quickly, as in the distributions in figure 4.13. The important thing to recognize is that the tail of the expected temperature is "fat" on the side of the distribution you would rather it weren't—it's fat on the hot side. Technically, this is known as a right-skewed Gaussian. When many model realizations are grouped together, their peaks average at around 3°C, which would indicate that Earth's average temperature would rise 3°C if CO_2 concentration were doubled.

On the low side, the distribution cuts off at zero, and many realizations cut off before reaching zero. That means that no model predicts cooling, and it's just a matter of how much warming we should expect. A distribution like this shows there is a finite probability that temperature could rise much more than 3°C but with very little chance that it could rise much less than 3°C. The probability of very high temperature changes does diminish, but very slowly. Although 3°C is the mean value, the shape of the distribution tells us that temperatures are likely to be higher than the mean rather than lower. Technically, for a regular Gaussian the mean and median value are identical, but for a skewed Gaussian they separate and the median (the most likely value) moves in the direction of the tail. The usual intuition we would like to follow is that the future will lie about midway between the two extremes, as in figure 4.12, but that is not correct—it is more likely to be closer to the darkened dot on the figure.

5

EARTH'S RESPONSES
TO CLIMATE CHANGE

SEA LEVEL CHANGES

Steric Sea Level Rise

In principle, one of the simplest calculations to make is the effect a new global temperature field would have on sea level. Imagine filling a container so the water level is one inch from the top; then heat the water. Heating will cause the water to expand (assuming the container itself does not change size at the same time and no water evaporates), and the water will come closer to the rim. The thermal expansion coefficient of water, usually written as α, is well established from experiments, so the prediction of sea level rise should be fairly straightforward. The shape of the ocean basins that contain seawater is also well known, and their shapes are not expected to change significantly. This is referred to as *steric* or *thermosteric* sea level rise, and it is associated with thermal expansion (and to a lesser degree to changes in salinity). Steric changes will also affect inland bodies of water, such as the Great Lakes in North America and Lake Baikal in Russia, Earth's largest lake by volume.

A number of factors make this calculation less simple than it might seem. One factor is that the entire ocean depth does not heat up all at once and expand as a single unit when heated from above. From the discussion of the thermocline in connection with ENSO (see chapter 4), we know that the upper parts of the ocean are much warmer and better mixed than the deeper parts. Heat introduced into the ocean from a warmer atmosphere takes many years to penetrate into the deep ocean, so the immediate response to warming will be primarily in the upper ocean. Furthermore, the thermal expansion coefficient α is itself temperature and salinity dependent. It increases as water temperature increases, so for the same rise in temperature, warm water expands more than cold water.

Like temperature rise, a family of sea level rise predictions is associated with a scenario and a model. Figure 5.1 shows a suite

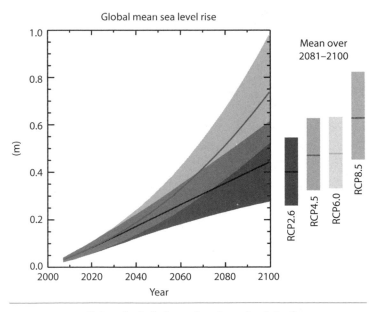

FIGURE 5.1 Suite of calculations of steric sea level rise for a range of different RCP scenarios.

Source: IPCC TS.5.4, 2007 IPCC Fifth Assessment report, Technical summary. Modified by http://www.easterbrook.ca/steve/2013/02/how-big-is-the-climate-change-deficit/.

of Intergovernmental Panel on Climate Change (IPCC) predictions for steric sea level rise derived in this manner.

A calculation of this sort has a number of shortcomings, and it provides the most conservative estimate of sea level rise. First, similar to average temperature, it does not include information on critical regional and local effects. The global oceans are affected by winds and internal dynamics, and some regions will experience greater sea level changes than others. These factors will surely change as Earth warms. Figure 5.2 shows recent observations of sea level changes, which are clearly strongest in the western Pacific. This is the area where sea surface temperatures and sea surface heights are typical highest due to the effect of the trade winds (see chapter 3). As might be expected, sea levels change with

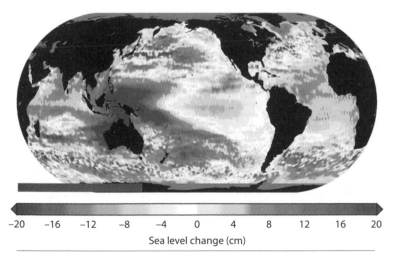

−20 −16 −12 −8 −4 0 4 8 12 16 20
Sea level change (cm)

FIGURE 5.2 Recent sea level increases obtained from NOAA satellite measurements. Darker areas have greater level of change. The area of strongest change is in the western Pacific where sea surface temperature is typically very high.

Source: NOAA Laboratory for Satellite Altimetry.

the ENSO cycle as winds vary in strength and direction from El Niño to La Niña to neutral years.

Second, this simple calculation is known to underpredict sea level changes that have occurred historically; that is, it is unable to simulate past sea level rise when temperature is also known. By correlating past temperature and sea level changes and using IPCC scenarios and models, a relationship can be established that gives estimates of sea level rise almost 3 times greater than simpler calculations.[1]

Contribution from Glaciers and Ice Sheets

A major source of uncertainty has arisen in understanding sea level rise due to the contribution of melting glaciers and ice sheets. First, it is important to distinguish glaciers from ice sheets. Glaciers are rivers of ice most often found in high mountain areas. They even exist in mountain ranges at the equator. Although glaciers are melting at a rapid pace, their contribution to sea level rise is not very significant. The much greater influence comes from the ice sheets that cover Antarctica and extend into the oceans, and the Greenland ice sheet that all but covers Greenland—these are called *grounded ice sheets*. If both ice sheets melted completely, it would cause tens of meters of sea level rise. However, even if warming is very strong, the complete melting of these huge ice sheets would take hundreds of years. Greenland's ice flows into the adjacent ocean from ice sheets via *outlet glaciers*. Sea level rise that takes place by this mechanism is called *eustatic* sea level rise. This melting will not alter the level of most inland water bodies.

Although it would take hundreds of years for the ice sheets to fully melt, we can calculate the contributions of their partial melting on sea level rise in the near future. Only ice that is currently grounded will contribute to sea level rise upon melting; floating sea ice ought simply to melt without changing the total volume of water (for example, an ice cube floating in a glass of water will not cause the glass to overflow if it melts). Recently sea ice volume has been reducing more rapidly than any of the IPCC models would suggest, and the volume of the massive ice sheet of Greenland is also declining much faster than simple calculations might predict.

The mechanisms that promote melting beyond simple solar heating of a static body are referred to as dynamic effects. Grounded ice sheets are often riven with deeply penetrating fissures that can extend to their base, and water from surface melting can enter these fissures and be carried to great depths, increasing the melting of ice not directly exposed to solar radiation. This water lubricates the interface between the ice and the rock below, allowing outlet glaciers to slip more easily toward the oceans. This, of course, does not explain the rapid melting of sea ice that might be caused by warmer ocean currents promoting melting from beneath.

The loss of grounded ice sheets has other effects. The ice sheets are so massive that they actually weigh down the crust of Earth, pushing it into Earth's mantle. If the ice could be instantaneously removed from Greenland, much of the central part of the land would be below sea level, perhaps forming a huge lake. As ice sheets melt, the outermost regions shrink and retreat toward the pole. As the weight is lifted, the crust rebounds by rising slowly upward (figure 5.3). This phenomenon is described as *isostatic rebound*, the process by which the crust and the mantle achieve equilibrium. The elevation of the crust is due to its mass only, and parts of Scandinavia are still rising following the last major glacial period. Those places may not experience sea level rise but instead experience a falling sea level, producing new coastal land areas. Currently the rate of uplift exceeds the rate of sea level rise.

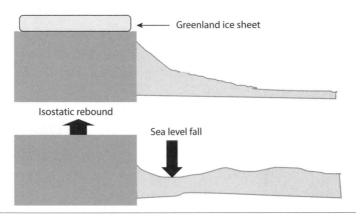

FIGURE 5.3 Stylized effect of the complete melting of the Greenland ice sheet.

A second factor is the gravitational effect of ice sheets on the surrounding oceans. The ice sheets are so massive that they attract the surrounding ocean water toward them, as the rough illustration in the upper part of figure 5.3 shows. When the ice sheets melt, their gravitational effect is removed and sea level in adjacent oceans experience a fall, not a rise. Figure 5.4 shows the effect on sea level if the West Antarctic ice sheet were to fully melt.

The effects of sea level rise are often portrayed as being simple to anticipate—they will move coastlines inland and submerge low-lying islands. Regions where the land surface slopes gently toward the sea will be most affected, and this includes delta regions such as the Mississippi Gulf Coast and the Ganges Delta, which makes up much of the country of Bangladesh. There are sixty-seven delta regions around the world, and they are extremely fertile farmlands.

Delta regions are very dynamic in nature, and their extent and morphology is a combination of sea level and sediment input from the large rivers that create the delta. Under most conditions, deltas grow in size into the sea. The source rivers are as important

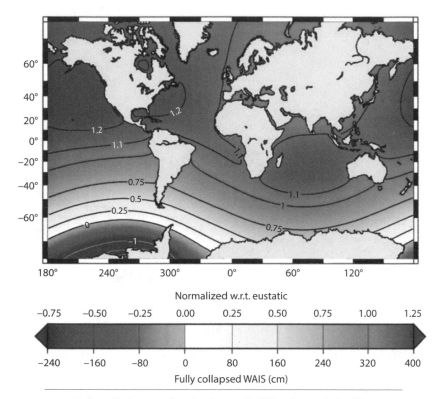

Normalized w.r.t. eustatic

| -0.75 | -0.50 | -0.25 | 0.00 | 0.25 | 0.50 | 0.75 | 1.00 | 1.25 |

| -240 | -160 | -80 | 0 | 80 | 160 | 240 | 320 | 400 |

Fully collapsed WAIS (cm)

FIGURE 5.4 Estimates of sea level were the West Antarctic Ice Sheet to fully to melt completely.

Source: Jonathan L. Bamber, Riccardo E. M. Riva, Bert L. A. Vermeersen, and Anne M. LeBrocq, "Reassessment of the Potential Sea-Level Rise from a Collapse of the West Antarctic Ice Sheet," *Science* 324, no. 5929 (2009): 901–903. 10.1126/science.1169335.

to the health of a delta as are the adjacent oceans. Construction of dams and levees that restrict the flow of sediment to the delta have a strongly detrimental effect. In Bangladesh, a rise of sea level by a meter, which is predicted under some scenarios at the end of the century, would flood a vast area of productive farmland and displace millions of people (figure 5.5). The capital, Dhaka, escapes this fate.

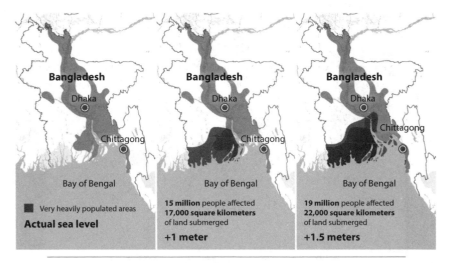

FIGURE 5.5 Impact of sea level rise in Bangladesh.

Source: GRID-Arendal, "Impact of Sea Level Rise in Bangladesh,"
http://www.grida.no/resources/5648.

The great majority of the world's major cities are located on coasts, often at the mouths of major rivers that provided inland access to farming areas in their earliest phases of growth. New York City on the Hudson River is a quintessential example of a coastal city that quickly became a major port, and it is one of the world's great centers of economic and cultural activity. Scores of major cities throughout the world, such London on the Thames and Cairo on the Nile, are in similar settings and will be strongly affected by even modest amounts of sea level rise.

FREQUENCY OF OCCURRENCE
OF TROPICAL CYCLONES

It is often stated that intense storm activity will increase in both frequency and strength in a warmer world. That intuition is largely based on the reasonable assumption that a warmer world would have a more energetic atmosphere and, hence, more vigorous atmospheric disturbances. However, clear observations show that the occurrence of Atlantic hurricanes is reduced in El Niño years (see figure 3.14), when the ocean is warmest, which contradicts this intuition.

Tropical Cyclone Formation

Tropical storms are categorized on the *Saffir-Simpson scale*, which ranges from 1 to 5, with the categories divided according to maximum sustained winds. "Sustained winds" means winds that persist at a given speed for more than a minute. Wind speeds have occasionally reached more than 200 mph, which exceeds the range of Saffir-Simpson's Category 5. A superior measure is the Power Dissipation Index (PDI), which aggregates intensity, and duration of a cyclone in one number.

Tropical storms are defined as tropical cyclones (TCs) if they have sustained winds greater than 64 knots (74 miles per hour) and display the characteristic rotating structure.[2] Tropical cyclone is the term used in scientific literature for this general class of meteorological event, although they are also referred to as hurricanes and typhoons, depending on the world's ocean basin where they occur. In other parts of the world, different strength scales, different means of sustained winds, and different naming schemes are used.

TCs generally evolve from irregular storm disturbances in the tropics. A number of scattered, relatively small storm systems may coalesce, evolve into a single large system, rotate under the influence of the Coriolis effect (provided they are not within 5° of the equator; see box 2.1), and may then build into a cyclone. They are not classified as cyclones and christened with a name

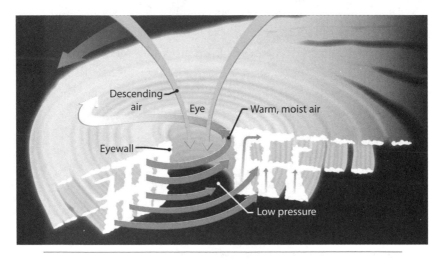

FIGURE 5.6 The basic elements of a tropical cyclone, with features noted.

Source: NASA Precipitation Measurement Missions, "How Do Hurricanes Form?,"
https://pmm.nasa.gov/education/articles/how-do-hurricanes-form.

until they develop these characteristics: rotation, a central eye of calm wind conditions (sea state may be high), intense eyewall winds, and very low atmospheric pressure at the center (less than 900 hPa—*hectopascals*) with surrounding sheets of heavy rain (figure 5.6). At the point of formation of a cyclone, there in no classical *weather front* such as is associated with a typical storm system. Technically this means a cyclone is nonbaroclinic; *baroclinic* cyclones are extratropical rotating storms initiated by a contrast between bodies of warm and cold air and are associated with a weather front. The transition from disorganized storms to an organized rotating cyclonic storm is not inevitable, and it requires a number of conditions to be present. Even when all conditions are in place for a cyclone to form, that does not ensure that one will form.

The conditions needed for formation of a cyclone follow:[3]

Warm ocean waters. This provides the moisture that is driven aloft
as water vapor and condenses to create intense rainfall characteristic of cyclones. Warm waters must be present for 100
m or so beneath the surface to provide sufficient continuous
fuel for the cyclone to develop. The heat released as water
vapor changes to moisture (the latent heat of condensation)
and further drives cyclone dynamics. Cyclones can be thought
of as heat engines, deriving energy from very warm ocean
waters and the condensation of water vapor. TCs rapidly lose
strength if they make landfall because their source of warm
water is cut off. They also lose strength as their path moves
into colder water environments.

Rotational force. This is provided by the Coriolis effect (see box 2.1).
The low pressure vortex drives moister air into the center
of the cyclone, maintaining a supply of energy for the
disturbance. The Coriolis effect is also responsible for the
typically accurate overall shape of a cyclone's path.

Low vertical wind shear. This refers to the difference in wind
speed between the sea surface and in the upper atmosphere
at the maximum height of the cyclone's disturbance. High
wind shear inhibits the vertical development of a cyclone (see
chapter 4) and may cause a cyclone to dissipate if strong wind
shear is encountered after formation.

Global Distribution of Cyclone Nucleation Regions and Track

The global distribution of tropical cyclone tracks is shown in
figure 5.7. The commonly stated intuition that TCs will increase in

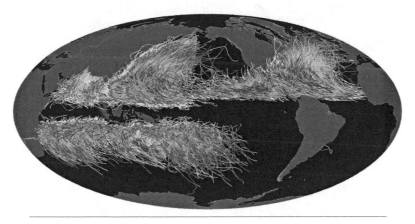

FIGURE 5.7 The global distribution of cyclone tracks dating back to 1850, compiled by NOAA. The most intense cyclones occur in the western Pacific. Cyclones that form there are called typhoons.

Source: NOAA National Centers for Environmental Information.

frequency and severity in a warmer world is based on the first of the conditions just described. It is expected that oceans will warm in the future, providing more of the drivers of TCs. It should, at the same time, increase the distance they travel as the tropics expand. But to understand how the incidence of cyclones might change in a warmer world, all three conditions need to be considered.

Rotational dynamics will not change enough to affect the formation of cyclones. Cyclones do not form within 5° of the equator because the Coriolis effect reduces to zero at the equator (strictly, it is the horizontal component of the Coriolis vector that is zero, the vertical component is actually strongest at the equator; see box 2.1). In a warmer world, cyclones will not form any closer to the equator than they do today because the rotational speed of Earth, which governs the Coriolis effect, will not change.

The global map also shows several other important features. TCs are much less frequently observed in the southern than in the northern hemisphere and are absent from large parts of the oceans in both hemispheres. South America has experienced only one TC in the last 150 years; Madagascar does experience TCs, but the continent of Africa is largely spared; the tropical South Atlantic, and much of the South Pacific, experiences very few TCs. TCs typically make landfall on east-facing coastlines. The question, then, is "How might this pattern be expected to change as climate change progresses?"

There is no reason to think that west-facing coastlines will become subject to TCs in the future. All cyclones are now and will continue to be initiated in the eastern region of oceans and move west under the influence of Earth's rotation. In the Atlantic and northwest Pacific, they typically have a hook-shaped track, ensuring that they stay over the ocean until they meet cooler water conditions in the north and dissipate (see figure 5.7).

TCs that begin off the west coast of Mexico do not follow the same sort of hook-shaped path, and they do not extend north very far. As we saw in discussing the role of oceans in forming climate zones (see chapter 3), ocean waters on the west coast of the United States are much cooler than those on the east coast at the same latitude, which is due to the influence of major circulation gyres in both oceans. Cool water strongly inhibits the growth of hurricanes and restricts most of these eastern Pacific hurricanes from traveling much further than 35° N. Some have made landfall in Baja California, but this is quite rare. A small number, having traveled more than halfway across the Pacific, also hook up to the north, but they too are uncommon.

The lack of hurricanes in the tropical south Atlantic is due to lower sea surface water temperature than in the north Atlantic and because this is an area of typically strong wind shear. The formation of hurricanes in the north Atlantic basin is strongly influenced by storm disturbances that occur in the Sudanese mountains of the Sahara and are carried west by so-called African easterly waves. About one in ten of these disturbances is carried across the west coast of Africa, where a few re-form into hurricanes over the ocean. In many cases, the tracks of Atlantic hurricanes initiate just off the west coast of Africa (see figure 5.7). Early season hurricanes initiate further west in the Atlantic Ocean. At that time of year, hurricanes may initiate in or very close to the Caribbean or the Gulf of Mexico, giving these regions little time to prepare. There is no equivalent to the meteorological conditions suitable for cyclone formation in South America or the northern South Atlantic.

The north and south extent of TCs is primarily governed by the temperature of the shallowest part of the ocean. In a warmer world, the region of warm ocean waters is expected to spread farther toward the poles. That is, the Hadley cell along with the tropics are expected to expand, and the extent of warm conditions in the upper ocean in both hemispheres should expand along with it. Because the energy source for cyclones is warm water, it is reasonable to suppose that TCs should be experienced farther north in the northern hemisphere than in the past and become more active in the southern hemisphere.

It is important to recognize from this discussion that two factors are in play: one is the area of formation of TCs, and the other is the track they follow. Hurricanes typically intensify as they track across the ocean, so the final strength achieved may have little to do with specific conditions where they begin. In assessing the changes in Atlantic hurricanes, the changing conditions in Saharan Africa become important. In thinking about how TCs may be influenced by warming, factors that cause them to both initiate and develop need to be taken into account.

If TCs are influenced by climate change, the record of past hurricanes could show an influence because there has been a warming trend of around 1°C in the last fifty years or so. This is not as straightforward as it might seem. NOAA has assembled an extensive set of observations of hurricane track and strength information through the middle of the nineteenth century. This data set, along with global sea surface temperature (SST) records, has been extensively studied to address the question of whether TC frequency has changed in recent decades in response to rising SSTs. The following discussion draws heavily on Knutson et al.'s[4] detailed assessment of tropical cyclones and climate change.

The results are equivocal in good part because the record is incomplete and unreliable in the time period before satellite observations. This is especially true of cyclones that stay at sea. Knowledge of those cyclones relies on opportunistic reports from ships in the area, and ships make great efforts to avoid cyclone weather. Commercial and military ships do not cover the oceans in a uniform pattern, and that pattern has changed considerably over time. The early record underestimates the number of cyclones that can be associated with rising SSTs because there is little doubt that many unrecorded TCs remained offshore. When adjusted for this issue, there is no discernible increase in Atlantic TCs. There is a distinct upward trend in cyclone frequency beginning around 1995, but when set against the adjusted long-term trend, this small trend visible in the records is not statistically significant. The conclusion is that there is "no compelling evidence for a substantial greenhouse warming-induced long-term increase [. . . .] in hurricane frequency."[5] Even less can be said about tropical cyclone frequency trends in other ocean basins.

What is observed, at least in the North Atlantic, is an increase in hurricane strength as measured by the PDI (figure 5.8). The trend for the North Atlantic looks quite clear, but the eastern North Pacific trend seems to display a decrease.

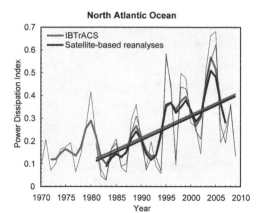

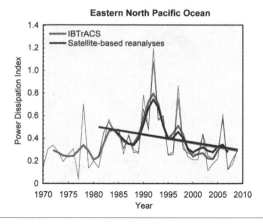

FIGURE 5.8 Change in PDI in the Atlantic and Pacific basins since 1972.

Source: U.S. Global Change Research Program, "Changes in Hurricanes,"
National Climate Assessment 2014, https://nca2014.globalchange.gov/report/our
-changing-climate/changes-hurricanes#narrative-page-16571.

Another approach involves computer modeling. Most large-scale global climate models predict both a significant warming of Atlantic SSTs and corresponding increases in upper tropospheric temperature and vertical wind shear. The first of these would stimulate cyclone activity; the second two would inhibit cyclone formation. Regional downscaled models that simulate a portion of the global system at finer resolution are required to discern which factors might be most important.

The result of these model studies is fairly inconclusive: "there is little evidence from current dynamical models that twenty-first-century climate warming will lead to a large (~30 percent) increase in tropical storm numbers, or PDI in the Atlantic."[6] In fact, the modeling suggests a substantial decrease in the number of Atlantic hurricanes. That said, the expectation from model studies is for stronger rainfall in hurricane rain sheets and for the frequency of occurrence of the most intense storms (Category 4 and 5 on the Saffir-Simpson scale) to increase despite an overall reduction in total numbers, which seems to be supported by figure 5.8. In brief, this means that it will become harder for hurricanes to form because of enhanced wind shear; but once formed, there is an increased possibility for hurricanes to build great strength due to warmer ocean temperatures.

These model studies indicate that the region of the ocean where hurricanes attain their maximum strength will move northward as expected. Given these features of projected hurricanes—fewer in overall number, but more of those becoming very intense—the future destructiveness (measured in PDI) of Atlantic hurricanes may increase by around 30 percent. That is, an increase in the number of high-category storms will outweigh the reduction in total storm numbers.

In a warmer world with a more energetic atmosphere, the consequence for the frequency and strength of warm world tropical cyclones is not very intuitive. Cyclones require alignment of a suite of factors to form and develop. Some of these factors may be enhanced, others may be diminished. Knutson et al.'s conclusions follow:

Detection and attribution. It remains uncertain whether past changes in any tropical cyclone activity (frequency, intensity, rainfall, etc.) exceed the variability expected through natural causes, after accounting for changes over time in observing capabilities.

Frequency. It is likely that the global frequency of tropical cyclones will either decrease or remain essentially unchanged due to greenhouse warming. We have very low confidence in projected changes in individual basins. Current models project changes ranging from −6 to −34 percent globally, and up to ±50 percent or more in individual basins by the late twenty-first century.

Intensity. Some increase in mean tropical cyclone maximum wind speed is likely (+2 to +11 percent globally) with projected twenty-first-century warming, although increases may not occur in all tropical regions. The frequency of the most intense (rare/high-impact) storms will more likely than not increase by a substantially larger percentage in some basins.

Rainfall. Rainfall rates are likely to increase. The projected magnitude is on the order of +20 percent within 100 kilometers of the tropical cyclone center.

Genesis, tracks, duration, and surge flooding. We have low confidence in projected changes in genesis location, tracks, duration, or areas of impact. Existing model projections do not show dramatic large-scale changes in these features. The vulnerability of coastal regions to storm surge flooding is expected to increase with future sea level rise and coastal development, although this vulnerability will also depend on future storm characteristics.[7]

THE FUTURE OF CLIMATE VARIABILITY

Two issues need to be considered when thinking of climate in the future. One is absolute temperature changes and the consequences in terms of such things as sea level rise, hurricanes, etc. The second is the variability of climate and its consequences. We learned quite a bit about variability in the discussion of ENSO. The phase of ENSO (El Niño or La Niña) can bring destructive droughts one year and equally disastrous flooding the next year. ENSO and other oscillations are referred to under the general term "climate variability." Until recently, climate models have not provided much information about climate variability because the focus has been, understandably, on long-term climate change.

A general answer to the question of future climate variation has come from the examination of a suite of models in which the variability, in broad terms, has been the focus of study. The results are shown in figure 5.9, and these results are not uplifting from a sustainable development perspective.

Each of the three letter notations is the location on the graph of a particular country. The horizontal line separates countries expected to experience increased climate variability in the future (above the line) from countries expected to experience reduced variability. The countries are also categorized by *gross domestic product* (GDP) per person: poorest countries to the left, richer countries to the right. With a few exceptions, poorer countries should expect greater variability and rich countries should experience less variability.

GDP wealth has nothing directly to do with physical climate variability. What is causing this distribution is that the tropics, where most of the poorest people live today, is expected to have a more variable climate in the future, whereas temperate countries, where global wealth is concentrated, will have diminished variability.

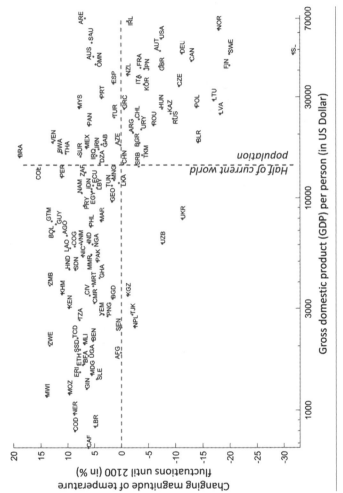

FIGURE 5.9 Future climate variability as a function of GDP wealth.

Source: Sebastian Bathiany, Wageningen University.

From the study of ENSO (chapter 3), we know that climate variations have direct effects on agriculture and disease in the poorest places, so the future may be even more challenging in currently poor regions of the world.

COMMITTED WARMING

The concept of "committed warming" has been advanced to suggest that no matter what society might do to stabilize or reduce GHG emissions, the planet is destined to see a rise in temperature due to what is described as a momentum in the climate system because of the long lifetime of some GHGs. Carbon dioxide molecules, for instance, remain in the atmosphere on average for more than one hundred years (see chapter 4). It is important to be accurate about what is meant by this idea, and that brings us back to the discussion of emissions versus concentrations, which was part of the discussion on climate models. Figure 5.10 makes the point.

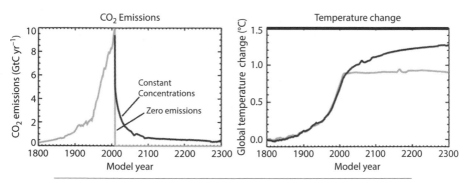

FIGURE 5.10 Computed consequences of bringing CO_2 emissions to zero.

Source: IPCC TS.5.4, 2007 IPCC Fifth Assessment report, Technical summary. Modified by http://www.easterbrook.ca/steve/2013/02/how-big-is-the -climate-change-deficit/.

If CO_2 emissions were to instantly go to zero at some point in time (shown as 2000 in figure 5.10, but the date is not relevant), as opposed to having emissions drop to a point where the concentration remained constant (the exact concentration level doesn't matter for this discussion), there are two quite different consequences. If emissions were reduced such that the concentration of CO_2 is stabilized, temperature will continue to rise for a long period due to the momentum effects of GHG longevity. However, if emissions are brought to absolutely zero—no emissions at all—GHG concentration will drop immediately because of the ability of the oceans to absorb CO_2.

The ocean is far from saturated with CO_2. In fact, a fall in temperature is possible. The total amount of CO_2 will diminish while the concentration in the ocean increases. The only factor that has a continued effect is the ocean because its heat capacity is large. The ocean would lose heat much more slowly than the atmosphere, causing the sea level rise to continue for quite some time before it begins to recover. The competition between the two scenarios will govern how long it will take for temperature to decline—and that may be hundreds of years. The important thing to note is that if emissions were to go to zero—and only if they were to go to zero—warming would not continue.

6

CLIMATE AND SUSTAINABLE
DEVELOPMENT

Summary and Closing Remarks

FEW topics are raised more often and have generated more contentious policy debates than that of climate change. Among the scientific community there is far less, if any, debate on this topic. Three critical questions are often asked about climate change, but the greatest controversy arises around the third question:

1. Will the climate of the future be different from today's, and how different will it be?
2. What might the consequences of a different climate be for people of the future and their sustainable development?
3. If it is determined that the consequences are bad, even if only for some, is it possible to stop the climate from changing (mitigation) or learn to live with the new climate regime (adaptation)?

In this book, I have focused most closely on the first of these questions and have provided some answers drawn from a very large body of current knowledge. Addressing this question can help us understand how climate might be altered by human activity.

If the factors that control Earth's average temperature are beyond human intervention, as some believe, then changing the climate is beyond our capability. To answer the question, we need a solid understanding of how the climate system functions. Early in this book I presented a simple representation of Earth, and we learned that changing the climate is well within our capacity because the composition of the atmosphere is critical to determining its near-surface temperature. Indeed, the climate of the future will be different if anthropogenic greenhouse gas (GHG) emissions are not stabilized and all but eliminated.

SUMMARY

The greenhouse effect was discovered by working through what seemed like a simple problem—determining why Earth has the average temperature it does today. No new physics or math are required to address this question. The major elements of the greenhouse effect were known well over one hundred years ago, long before computer simulations of Earth systems were possible, and well before any concerns were raised about global warming. The greenhouse effect is not a theory erected to explain observed warming trends; the greenhouse effect was operating well before these trends were first recognized and an explanation was sought.

In our crude first approximation, Earth is a static, one-dimensional object—it has no shape—there are no ocean currents (no ocean in fact), no winds, and no atmosphere except clouds. That is not a very accurate description of Earth, so it is perhaps no surprise that an analysis based on this model of Earth gives the wrong answer. Refining our depiction of Earth included a slightly more complete description of the atmosphere, although it was still one-dimensional. That led to a better answer, and to an appreciation of the greenhouse effect.

The greenhouse effect is explained as a quantum mechanics phenomenon. The atmosphere is a thin gas that includes molecules composed of three or more weakly bonded atoms. Long wavelength electromagnetic radiation emitted by Earth causes these molecules to vibrate, absorbing and reemitting a fraction of their energy. That energy causes a warming of the atmosphere. The energy is, in part, returned to Earth, causing it to be warmer than it otherwise would be. Tightly bonded two-atom molecules such as oxygen and nitrogen are in much greater abundance than other gases, but they do not participate in the greenhouse effect because they do not vibrate in response to long wavelength radiation.

I then made a second approximation in which Earth is modeled as a sphere, rotates on its axis, and has a temperature field that varies from equator to poles. The addition of these two factors can be analyzed using classical Newtonian physics, and this analysis provides answers to many aspects of global wind patterns, distribution of rainfall, and the location of deserts, rain forests, and other major features affecting the general circulation of the atmosphere. This leads to a depiction of the atmosphere with three large circulatory systems in each hemisphere.

To be confident that climate can be changed by anthropogenic forcing requires two things:

1. We know the basics of how the climate system works, and
2. Climate factors influenced by human activity may be distinguished from the natural variations that are an integral part of the climate system.

Natural variations are expressed in several ways. Milankovitch cycles are the result of long-term orbital variations of the

Earth-Sun system—the change in tilt and direction of the axis of rotation, and the shape of Earth's orbit around the Sun. They occur in periods from twenty thousand to more than one hundred thousand years. Human activity has no part in determining these changes, and these natural variations have given rise to glacial and interglacial cycles for at least five million years.

On shorter time scales are quasi-periodic oscillations of a few years (El Niño Southern Oscillation [ENSO]) to decades (Pacific Decadal Oscillation). The mechanism of ENSO is now quite well understood, and this has provided a degree of predictability useful in understanding agriculture and vector-borne disease management in the tropics, where the poorest people live and where development has proven to be the most difficult. The mechanism of other oscillations is not understood sufficiently to provide useful predictions. The ENSO phenomenon was discussed in some detail, emphasizing the importance of atmosphere-ocean feedback interactions. The methodology and challenge of ENSO prediction was also explained, with an assessment of the limits and uncertainties involved. Although generated in certain regions, these oscillations affect much larger parts of the planet. For instance, El Niño, the warm phase of ENSO, has impacts as far away as the Atlantic Ocean, where it suppresses the development of hurricanes by intensifying upper-level winds that cause wind shear. When the North Atlantic Oscillation remains in a positive state it appears to influence the frequency of El Niño events, making them more common than La Nina events.

Human activity is not believed to have a direct or significant influence on these oscillations, in part because there is evidence from indicators of past climates that ENSO, in particular, was operating much as it does today well before humans inhabited Earth. The oscillations arise from interactions between the oceans and the atmosphere that are completely natural in origin.

No aspect of human activity could interrupt this cycling. However, if the average temperature of Earth increased significantly due to human activities, it is plausible that this could interrupt ENSO dynamics, possibly by causing the warm phase (El Niño) to be more prevalent along with its global teleconnections.

Having achieved this level of description of the climate system, and gained some insight into climate prediction problems, the topic of determining future global climate states was investigated. The input parameters and basic functioning of climate models were described along with the critical role GHG emission scenarios play in the uncertainties of their outcomes. Of the many consequences an altered temperature field might bring to the planet's systems, two were discussed in some detail—sea level change and tropical cyclone frequency and strength—and it was found that simple intuition is not always rewarded when considering the consequences of temperature change. For instance, although it might seem plausible that a warmer planet would experience more tropical cyclones than it does today, the present state of understanding suggests that there will not be more, and their could even be fewer tropical cyclones. Of that smaller total, however, a greater proportion should reach very high strength and destructiveness. In addition, we now know that an increase in climate variability is most likely to occur in the tropics, with diminished variability in temperate zones.

The prediction of global climate change far into the future is an extraordinarily difficult problem, in large part because we have to predict the response to conditions that Earth has never before experienced. The question boils down to how sensitive Earth actually is to changes in the composition of the atmosphere. That sensitivity figure can be estimated in several ways, and current estimates suggest that a doubling of carbon dioxide concentration over preindustrial levels will lead to a temperature increase of about 3°C. The probability distributions of sensitivity values are in the form of a right-skewed Gaussian, so higher temperatures are more likely than temperatures below the mean.

Predicting the physical adjustments Earth may make in response to such large changes in temperature is even more difficult. Estimates of sea level rise must take into account far more than simple thermal expansion. At a level of much higher difficulty is predicting how human societies might respond to these changes, and that is an inquiry beyond the scope of this book.

CLIMATE CHANGE AND SUSTAINABLE DEVELOPMENT

Throughout the text, I have referred to the consequences for human welfare of climate phenomena. For instance, the way the phases of ENSO can influence the occurrence of malaria in Africa and their effects on staple crops was discussed in chapter 3. This section summarizes how a warmer climate might influence opportunities to gain welfare improvements.

Development encompasses the idea of an improvement in human welfare. Welfare itself can be measured in many ways and is most commonly reported as gross domestic product (GDP). Introduced in the late 1930s during the depression era, it is well recognized today that GDP is an inadequate measure of welfare for many reasons. One inadequacy is that GDP does not account for natural capital—the value of natural assets such as forests, rivers, and oceans or the state of pollution in cities. GDP emphasizes gains in manufactured capital, as was its intent. It cannot measure the inequality in welfare experienced in many countries. Nor can it measure the informal economy of work done or goods sold and traded in a largely cash economy with no taxes or avoided taxes. The economies in many poor countries have large unmeasured components of informal and illegal businesses as well as remittances from relocated people.

Even when measured as GDP, the state of development and the improvement of human welfare differ hugely around the world. *Sustainable development* takes on a different meaning in different parts of the world. In poor countries, the emphasis is on development itself because the state of welfare is so low. In richer countries—those that have experienced development—the emphasis is on sustainability. For these countries, we ask whether the current level of welfare can be maintained without causing (more) harm to the condition of the planet that could imperil the welfare of future generations. Global warming is a classic case of negligence of intergenerational welfare—the pathway to development that has brought prosperity to the present generation imperils generations to come.

The idea that a different global climate regime could significantly affect the welfare of human societies is, in itself, not controversial. Changes in climate conditions in the past have modulated human welfare in dramatic ways. Droughts leading to severe water shortages, crop failure, and famine have caused millions of deaths and massive population displacements, and these conditions were common in many parts of the world through the middle of the twentieth century. It has been argued that periods of intense drought may have led to the complete collapse of entire ancient civilizations, such as that of the Mayans. Migration of people from drought-ridden rural areas, where agriculture collapsed, into city centers is thought to have increased urban stresses and discontent with governments and may have triggered conflicts during the period called the Arab Spring. Conflict, whether internal or cross-border, has been described by Paul Collier as "development in reverse" because it wastes human capital and national wealth.[1] It is also thought that famine resulting from extended periods of extreme drought events, if summed from the distant past, has taken more lives that any other type of natural disaster.

Most of these examples can be better thought of as societies attempting to cope with climate variability rather than climate change. Except for the most recent examples, they occurred in times when the average global temperature was neither appreciably increasing nor decreasing. From recent studies, we now know that climate variability is likely to increase in the future along with average temperature, but like everything else, this will not be uniform across the globe. The unhappy outcome seems to be that variability will increase most in the tropics, where the majority of poor countries are located, and diminish in richer temperate zone countries. A wildly varying climate may be more difficult to cope with than a somewhat warmer but stable condition.

Today sophisticated warning systems monitor conditions in drought-prone areas of the world from satellites and other systems. One such endeavor is the Famine Early Warning System network created by USAID to monitor and reduce the possibility of famine conditions, which and are restricted entirely to poorer parts of the world. In a world where the average temperature is greater than today, it is expected that extreme heat events, even if as brief as a few days, will become more common and more extreme.

From a sustainable development perspective, drought conditions affect the welfare of poor countries the most and are a strong inhibitor of their development. Famine deaths are much less common today, but they do still occur and are almost exclusively in the worlds' poorest regions. These conditions are often aggravated when associated with civil conflict and fragile governments such as the current conflict in Yemen. Deaths may not be from starvation itself but from illnesses that take hold of weakened bodies, especially those of children. North Korea is thought to have experienced agricultural failures that led to starvation, and these failures may have had as much to do with bad policy decisions as climate conditions. Mass starvation in China during the Great Leap Forward (1958–1962) under Mao Zedong had a notable effect on the world's total population; this was the result of tragically bad agricultural policy decisions.

Drought conditions significantly disrupt the economic progress of poor countries because many are heavily dependent on agriculture for their economic welfare. In many poor countries, even at the best of times, crop harvests are very low when measured in bushels per hectare. The difference is almost tenfold compared to developed countries, and countries that depend most on agriculture have smaller agricultural outputs. The Green Revolution, which created massive increases in crop production and the end of famine in India, does not yet have an equivalent in the poorer parts of Africa where arable land is scarce and crops are very different from those in India.

The importance of agriculture in a development context is that a typical path to prosperity first involves the success of agriculture in supplying adequate food for society. What follows has often been the development of industries to produce goods for domestic needs and to obtain export earnings. This leads to a rise of service industries that today make up the bulk of the economies of many developed countries such as the United States, where the service sector accounts for a little more than 80 percent of GDP and agriculture less than 1 percent. The economic mix begins with a dominant agricultural base and evolves to a dominant service base, with manufacturing occupying a middle place. Agricultural production itself does not diminish, but its role in the aggregate economy does. As fragile as agriculture is today in many poorer countries, it could become more precarious still in a warmer, more climatically variable world, and economic progress would be threatened as a consequence.

In wealthy countries, drought usually does not have a serious impact on the aggregate economy because more wealth is derived from services and manufacturing, which are inherently more robust during climate fluctuations. Droughts in the American Southwest have been severe, yet they have had little to no effect on the aggregate U.S. economy. Nevertheless, drought can cause major losses to crop production in wealthy exporting countries, leading to sharp fluctuations in food prices globally, with the major impact in poorer regions that import much of their food needs. Puerto Rico, for instance, imports more than 90 percent of its food needs and has acquired enormous national debt, with eleven straight years of negative growth as a consequence. The Puerto Rican diaspora is huge. Today more Puerto Ricans live on the U.S. mainland than on the island, and the diaspora continues to grow. France and other European countries experienced heat waves in 2003 that caused more than 35,000 deaths, which temporarily disrupted their economies. In France the heat caused extensive crop losses, and mortality, primarily among the elderly, was greatest at 14,802 deaths.[2]

El Niño conditions can cause drought and flooding simultaneously. In India it can disrupt the monsoon rains critically needed for agriculture. Failure of the monsoon means massive crop losses that have historically led to starvation. Most climate models suggest an intensification of monsoon rains in the future, together with greater variability. More intense rains will surely lead to more flooding, but variability may be more difficult to manage as agricultural practices are built around the expectation of reliable monsoon rains. Most climate models suggest that the world will be wetter overall (due to enhanced evaporation from the oceans), with regions where rain is intense at present becoming wetter and dry places becoming dryer.

From a physiological perspective, healthy human beings can tolerate quite high temperatures without great suffering. There is ample evidence, though, that human productivity in hot climates is considerably diminished, even in regions where hot conditions are common and one might expect some adaptation to have taken place. Cold conditions limit human habitability much more than do hot conditions. Our species may be able to take the heat in a physiological sense, but it is a different matter altogether for the plethora of natural systems on which human survival depends. The great boreal forests followed the retreating Arctic ice sheet northward at the end of the last glacial maximum and presumably could do so again if the rate of retreat due to climate change is not very rapid. But the great majority of plants and animals humans use for food and clothing have much more restricted ranges, and some have quite small ranges. Many plant species that are critical food sources have been bred and optimized to grow in narrow climate conditions. In other words, most of our staple crops are highly specialized to conditions that will not exist in the future in their present location.

For plants to grow well, they need more than just a specific ambient temperature to which they are most suited. Plants also require adequate light, rainfall, and appropriate soil type and quality. Rainfall must be available at critical times in a plant's growing cycle, and even the ambient temperature may be less important than the prevalence and timing of extreme temperatures—only a few very hot days can ruin a maize crop that had otherwise been growing well, and a brief cold spell can ruin many citrus crops. Current climate models suggest that climate variability will increase in the poorer/tropical parts of the world and decrease in temperate regions, making development even more difficult in poor countries and more reliable in developed countries.

Climate will not change uniformly around the world, nor will climate variability; some regions are expected to change much more than others. The Arctic and the Antarctic peninsular are showing the strongest evidence of change today, which is expected due to the ice albedo feedback effect. The hottest regions, despite warming less overall than the poles, might experience the greatest change in variability. These regions could well reach a state so hot and variable that it would be impossible to support humans. Rivers and lakes and soil may dry up, and farming that is difficult today may become all but impossible. Climate change in the Arctic is dramatically altering natural conditions, such as the extent of sea ice and altered animal habitat, but very few people inhabit these areas. The tropics are much more densely populated, and changes in the tropics are likely to be the most important from the perspective of human welfare.

Sea level is sure to rise in most places, although also very unevenly, and sea level may even fall in a few places. Coastal regions in both rich and poor nations will submerge and become subject to storm damage if no protective barriers are erected. Islands will shrink in area, and some will submerge altogether so their residents will need to be relocated. Very few island states are wealthy. Their economies are restricted and usually revolve around tourism, aid from former colonizers, and offshore banking. Coastal cities are the hub of economic activity in most developed countries, and in many poorer countries also. Fertile soils in river deltas, such as the Mississippi, the Ganges-Brahmaputra, and the Irrawaddy of Myanmar, make them ideal for farming, but they are so close to sea level today that even a modest rise in sea level could inundate vast areas of farmland. These same regions and most east-facing coastal regions will experience a greater number of the most intense tropical cyclones, compounding the plight of people in coastal cities who will face the decision of relocating inland (or to higher elevation) or creating expensive protective infrastructure, which is well beyond the capacity of poorer countries.

It is widely thought that sea level changes, combined with a rising number of intense cyclones, present the greatest threats to human welfare that climate change might bring in many parts of the world, rich and poor. Cyclones cause massive damage to private and public buildings and infrastructure and affect economies in several ways. One way is the replacement cost of destroyed or damaged structures; the built capital losses figure is most often given in media reports soon after a cyclone. In comparing the damage to capital stocks in rich versus poor countries, we need to recognize that homes and other built infrastructure in poor countries may have a lower capital value than in rich countries but nevertheless be critical to the overall economy. Rich countries have reserves that permit them to recover quickly, whereas poor countries often show the physical damage of cyclones for many years. The loss of roads, port facilities, and bridges interrupt the flow of goods, and that may affect an economy more than the loss of capital stock. A cyclone disaster is at the very least a setback to development that affects poorer countries more than the rich, even though that may not be obvious in aggregate economic data.

Cyclones also cause unemployment and large population displacement when businesses are ruined. Human welfare cannot be advanced in refugee camps. Intense rainfall well inland of a cyclone's landfall causes considerable damage and fatalities as well. Cyclones in a warmer world are expected to be associated with more intense rainfall. (Hurricane Harvey, which hit Houston in 2017, may be a harbinger of this phenomenon.) The cost of renting a home often rises in disaster areas due to the scarcity of rental properties. Gasoline and food prices rise for the same reason. Often described as indirect effects, they can have a greater effect on an economy than direct capital losses.

Global warming can be thought of as causing an expansion of the tropics together with a shift of temperate zones toward the poles and a shrinking of polar regions. As the tropics expand, the range of tropical diseases such as malaria will also expand. Less than a century ago, malaria had a much greater range that it does today. Malaria was driven into its current range by the use of chemicals that are now banned. Malaria is a massive public heath issue in poor countries where health systems are weak, and it causes considerable loss of human capital. In the quest to eradicate malaria, climate change is generally considered a minor issue, but already vector-borne diseases are appearing in the southern states of the United States where they have never been seen before or were thought to be eradicated long ago. This may not impede development of the United States, but it is certainly a concern.

We can be fairly certain that the harms of climate change, whatever they may be, will not be felt equally. The very size and momentum of wealthy countries' economies, as well as their economic makeup, focused more on the service sector than agriculture and manufacturing, could make them inherently more robust to environmental changes. Most of the highly developed countries are in temperate zones (the Ferrel cell). Wealthy societies may be able to adapt in place or move their centers of economic activity. Poor countries already suffer much more from natural climate variations and associated extremes. The differential impacts of climate change may increase existing global welfare disparities. The irony—often stated in discussions of climate change and sustainable development—is that those who created the problem and have become wealthy through industrial emissions will be the most prepared to cope with its consequences, whereas those who are the least developed and have participated least in causing the problem will be the least able to cope with its consequences.

GLOSSARY

ABSORPTION SPECTRUM. The spectrum of electromagnetic radiation having passed through a substance.

AEROSOLS. A system of solid particles uniformly distributed in a finely divided state through a gas, usually air.

ALBEDO. Reflective power or reflectivity; specifically the fraction of incident radiation (such as light) reflected by a surface or body (such as the moon or a cloud). Usually measured as a percentage or decimal factor between 0.0 and 1.0.

ATMOSPHERE. The envelope of gases surrounding Earth or another planet.

AVERAGE RESIDENCE TIME. Also known as removal time, this is the average amount of time spent in a control volume by the particles of a fluid.

BAROCLINIC. A measure of the stratification in a fluid.

BJERKNES FEEDBACK. Interaction between the ocean and the atmosphere in the central equatorial Pacific in which the easterly surface wind stress enhances the zonal sea surface temperature gradient, which in turns amplifies the wind stress.

BLACKBODY. An idealized physical body that absorbs all incident electromagnetic radiation, regardless of frequency or angle of incidence. It is simultaneously a perfect emitter of radiation.

CAUSATION VS. CORRELATION. If there is a correlation between two measured quantities, we sometimes assume that the dependent variable changes solely because the independent variables change. However, correlations can be spurious and related to changes in an unmeasured quantity, in which case the correlation does not imply direct cause and effect.

CETERIS PARIBUS. With all other conditions remaining the same.

CHAOTIC PROPERTIES. A branch of mathematics focused on the behavior of linear dynamical systems that are highly sensitive to initial conditions. Earth's climate is an example.

CLIMATE SENSITIVITY. The factor relating change in temperature for a given change in radiative forcing.

CLIMATOLOGICAL EQUATOR. The point on Earth where the direct rays of the Sun strike.

CLOUD RADIATIVE FEEDBACK. Clouds are both reflective to energy received from above and absorbing to energy received from below. Changes in Earth's temperature will give rise to changes that can enhance or diminish these properties.

CORIOLIS EFFECT. The effect of Earth's rotation on bodies not attached to Earth but moving N-S or S-N across Earth's surface.

CRYOSPHERE. Those parts of Earth where water is in solid form.

CYCLOTHERM. Alternating fine stratigraphic sequences of marine and nonmarine sediments, sometimes interbedded with coal seams.

DOLDRUMS. A colloquial maritime expression referring to those parts of the Atlantic and Pacific oceans affected by a low pressure area around the equator where the prevailing winds are calm.

DYNAMIC MODELS VS. STATISTICAL MODELS. In mathematics, a dynamic system is one in which a function describes the time dependence of a point in a geometrical space. Examples include the mathematical models that describe the swinging of a clock pendulum, the flow of water in a pipe, and the number of fish each springtime in a lake. A statistical model is a class of mathematical model that embodies a set of assumptions concerning the generation of some sample data, and similar data from a larger population. A statistical model represents, often in considerably idealized form, the data-generating process.

ECCENTRICITY. Deviation of a planet's axis of rotation from normal to the plane of the ecliptic.

EEMIAN PERIOD. The interglacial period that began about 130,000 years ago and ended about 115,000 years ago. It is the most recent period before the present day in which temperature was similar to that of today.

EL NIÑO SOUTHERN OSCILLATION (ENSO). An oscillating weather phenomenon associated with a band of warm ocean water that develops in the central and east-central equatorial Pacific, including off the Pacific coast of South America. ENSO switches between a warm state, El Niño, and a cold state, La Niña, in periods between four and seven years.

ELECTROMAGNETIC SPECTRUM. The range of wavelengths exhibited by electromagnetic radiation. Classified by wavelength into radio wave, microwave, terahertz (or submillimeter) radiation, infrared, or visible region that is perceived as light, ultraviolet, X-rays, and gamma rays. Wavelength is determined by the temperature of the body emitting the radiation.

ELLIPTICITY. Having the characteristics of a regular oval shape, traced by a point moving in a plane so that the sum of its distances from two other points (the foci) is constant, or resulting when a cone is cut by an oblique plane that does not intersect the base. In climate science, it refers to the deviation from circular of Earth's orbit around the Sun; all the solar planets have this feature.

EUSTATIC. Relating to or characterized by worldwide change of sea level caused by changes in the solid Earth and the melting of glaciers.

EVAPOTRANSPIRATION. The sum of evaporation and transpiration from surface water bodies, plants, and soil.

FEEDBACK. The modification or control of a process or system by its results or effects.

GEOENGINEERING. The deliberate, large-scale manipulation of an environmental process that affects Earth's climate in an attempt to counteract the effects of global warming.

GLACIAL PERIOD. A period in Earth's history when polar and mountain ice sheets were unusually extensive across Earth's surface.

GLOBAL WARMING POTENTIAL (GWP). A relative measure of how much heat a greenhouse gas traps in the atmosphere. It compares the amount of heat trapped by a certain mass of the gas in question to the amount of heat trapped by a similar mass of carbon dioxide.

GREENHOUSE EFFECT. Trapping the Sun's warmth in a planet's lower atmosphere due to the greater transparency of the atmosphere to visible radiation from the Sun than to infrared radiation emitted from the planet's surface.

GREENHOUSE GAS (GHG). A gas in the atmosphere that contributes to the greenhouse effect by absorbing and reradiating long-wavelength infrared radiation.

GROSS DOMESTIC PRODUCT (GDP). The total value of everything produced by all the people and companies in the country.

GTC. Giga tonne of carbon or carbon equivalent.

GYRE. A spiral or vortex in ocean surface currents.

HECTOPASCALS (HPA). Unit used to measure pressures, equal to 100 pascals or 1 millibar.

HERTZ (HZ). The unit of frequency in the International System of Units and is defined as one cycle per second. It is the inverse of wavelength.

HIGGS BOSON. An elementary particle in the standard model of particle physics.

HORSE LATITUDES. Subtropical latitudes between 30 and 38 degrees both north and south where Earth's atmosphere is dominated by the subtropical high, an area of high pressure, which suppresses precipitation and cloud formation and has variable winds mixed with calm winds.

ICE ALBEDO FEEDBACK. A positive feedback climate process in which a change in the area of snow-covered land, ice caps, glaciers, and sea ice alters the albedo, which in turn alters the climate system.

INSOLATION. The amount of solar radiation reaching a given area; a constructed word from "incoming solar radiation."

INTERGLACIAL PERIOD. A geological interval of warmer global average temperature lasting thousands of years that separates consecutive glacial periods.

INTERTROPICAL CONVERGENCE ZONE (ITCZ). The area encircling Earth near the equator where the northeast and southeast trade winds converge.

ISOBARS. A line on a map connecting points having the same atmospheric pressure at a given time or, on average, over a given period.

JET STREAM. Region of very strong upper atmospheric winds. Each hemisphere has a polar jet and a subtropical jet. Wind speeds can exceed 200 mph.

LATENT HEAT. The quantity of heat absorbed or released by a substance undergoing a change of state, such as ice changing to water or water to steam, at constant temperature and pressure. Also called "heat of transformation."

MALARIA EARLY WARNING SYSTEMS (MEWS). International system that aids in prediction of malaria outbreaks. Organized under the World Health Organization.

MONSOON. A seasonal reversing wind accompanied by corresponding changes in precipitation, *or* seasonal changes in atmospheric circulation and precipitation associated with the asymmetric heating of land and sea that bring heavy rains.

NINO3.4. The average sea surface temperature anomaly in the region bounded by 5° N to 5° S, from 170° W to 120° W.

NORTH ATLANTIC OSCILLATION (NAO). A weather phenomenon in the North Atlantic Ocean caused by fluctuations in the difference of atmospheric pressure at sea level between the Icelandic low and the

Azores high that modulates the strength and direction of westerly winds and the location of storm tracks across the North Atlantic.

OBLIQUITY. The deviation from parallelism or perpendicularity; also the amount of such deviation. In climate science it refers to the deviation of the spin axis of Earth from being normal to the plane of the ecliptic.

OROGRAPHIC EFFECT. Applied to the rain or cloud caused by the effects of mountains on air streams that cross them.

OUTLET GLACIER. A glacier that connects an ice sheet to an adjacent ocean.

PACIFIC DECADAL OSCILLATION (PDO). A robust, recurring pattern of ocean-atmosphere climate variability centered over the midlatitude Pacific basin.

PARAMETRIZATION. The introduction of variables required for model calculations that cannot be based on observations but must be estimated from theory.

PLANCK FUNCTION. Formula that describes the spectral density of electromagnetic radiation emitted by a blackbody in thermal equilibrium at a given temperature.

PLANE OF THE ECLIPTIC. The plane in which Earth rotates around the Sun.

PLASMODIUM FALCIPARUM. A protozoan parasite, one of the species of Plasmodium that cause malaria in humans. It is transmitted by the female Anopheles mosquito.

POWER DISSIPATION INDEX (PDI). The sum of the maximum one-minute sustained wind speed cubed, at six hourly intervals, for all periods when the cyclone is at least at tropical storm strength.

PPB. Parts per billion is the number of units of mass of a contaminant per 1,000 million units of total mass. The unit used to measure the concentration of helium in the atmosphere and some other trace gases.

PPM. Parts per million. Usually describes the concentration of something in water or soil. One ppm is equivalent to 1 milligram of something per liter of water (mg/l) or 1 milligram of something per kilogram of soil (mg/kg). It is a common unit for concentration of carbon dioxide in the atmosphere.

PRECESSION. The slow movement of the axis of a spinning body around another axis due to a torque (such as gravitational influence) acting to change the direction of the first axis. It is seen in the circle slowly traced out by the pole of a spinning gyroscope. Earth experiences this motion, which is commonly referred to as the precession of the equinoxes.

RADIATIVE BALANCE. The relationship between the amount of energy reaching an object (or a portion of it) and the amount leaving the same object. In climate science, it refers to the energy balance at the top of the atmosphere, the troposphere.

RADIATIVE FORCING. The difference between radiation absorbed by Earth and energy radiated back to space.

RAIN SHADOW. The area on the leeward side of a mountain range that typically experiences dry conditions because rain has fallen on the windward side of the range due to orogenic uplift.

RESIDENCE TIME. The average length of time during which a substance, a portion of material, or an object is in a given location or condition, such as adsorption or suspension. In climate science, it refers to the amount of time a molecule of any gas remains in the atmosphere.

SAFFIR-SIMPSON SCALE. Classifies hurricanes (western hemisphere tropical cyclones that exceed the intensities of tropical depressions and tropical storms) into five categories distinguished by the intensities of their sustained winds.

SEA LEVEL PRESSURE (SLP). Pressure of the atmosphere measured at sea level, or reduced to sea level if measured at a different elevation.

SOLAR RADIATION MANAGEMENT (SRM). A type of climate geoengineering that seeks to reflect sunlight and thus reduce global warming. Proposed methods include increasing the planetary albedo, for example, by using stratospheric sulfate aerosols or the enhancement of cloud albedo.

SPHERICAL SPREADING. The decrease in energy level when a wave propagates away from a point source uniformly in all directions. Also known as "geometric spreading."

SST ANOMALY. Sea surface temperature anomalies, such as those associated with El Niño (La Niña), are a measure of the departure of temperature in the ocean's surface waters from normal conditions. One method is to use five consecutive three-month running means of sea surface temperature.

STEPPE. An eco-region in the montane grasslands and shrublands and temperate grasslands, savannas, and shrubland biomes, characterized by grassland plains without trees other than those near rivers and lakes.

STERIC. The amount of sea level change caused by expansion/contraction of the oceans due only to changes in temperature.

STRATOSPHERE. The second major layer of Earth's atmosphere, just above the troposphere, and below the mesosphere.

TELECONNECTIONS. Climate anomalies being related to each other at large distances (typically thousands of kilometers). Most commonly used to describe a weather pattern altered by ENSO.

THERMOCLINE. A steep, decreasing temperature gradient separating a uniformly warm region immediately below the surface that is heated by the Sun from a deep region of near uniform cold water beneath.

THERMOHALINE. The effects of salinity and temperature, especially with regard to large-scale ocean circulation.

TRANSIENT CLIMATE RESPONSE TO CUMULATIVE CARBON EMISSIONS. The factor, usually denoted λ, that assesses the temperature response of a climate model to a sudden doubling of the concentration of CO_2 in the atmosphere.

TROPOPAUSE. The boundary in Earth's atmosphere between the troposphere and the stratosphere.

TROPOSPHERE. The lowest layer of Earth's atmosphere, where nearly all weather conditions take place. It contains approximately 75 percent of the atmosphere's mass and 99 percent of the total mass of water vapor and aerosols.

TUNDRA. A type of biome where tree growth is hindered by low temperatures and short growing seasons. The term *tundra* comes from the Russian тундра from the Kildin Sami word *tūndâr* (uplands or treeless mountain tract).

VECTOR. A quantity having direction as well as magnitude, especially for determining the position of one point in space relative to another.

VEGETATION. Plants considered collectively, especially those found in a particular area or habitat.

WALKER CIRCULATION. In the tropics, the pattern of atmospheric circulation that includes near-surface winds blowing from east to west, and upper-level winds blowing from west to east. Circulation is broken into two cells during a strong El Niño.

WEATHER FRONT. A boundary separating two masses of air of different densities; it is the principal cause of meteorological phenomena outside the tropics. In surface weather analyses, fronts are depicted using various colored triangles and half-circles, depending on the type of front.

WHITTAKER BIOME DIAGRAM. A graphic in which the dependence of vegetation types globally depends on temperature and rainfall.

WIND SHEAR. Variation in wind velocity occurring along a direction at right angles to the wind's direction and tending to exert a turning force.

NOTES

INTRODUCTION: STRATEGY AND OUTLINE OF THE PRIMER

1. Bertrand Russell, *Skeptical Essays*, 2nd ed. (London: Routledge, 2004), 25.

1. WHY DOES EARTH HAVE THE CLIMATE IT DOES?

1. "Vertical" means at 90° to the plane of the ecliptic in which Earth revolves around the Sun.
2. Named for the nineteenth-century German physicist Heinrich Hertz.
3. Note that sound is an acoustic wave, not an electromagnetic wave, but it has similar wave characteristics.
4. Rays are commonly used to describe the propagation of seismic energy associated with earthquakes.

2. PRECIPITATION, WINDS, ATMOSPHERIC PRESSURE, AND THE ORIGIN OF CLIMATE ZONES

1. This is the most common story of the origin of the term *horse latitudes*, but there are at least two other stories dating from the sailing era.
2. The Coriolis effect was first noticed in the flight of cannon balls and in the way objects fell when dropped from very tall structures.
3. Evapotranspiration is water lost to the atmosphere from the ground surface and the transpiration of groundwater by plants whose roots tap the capillary fringe of the groundwater table. See U.S. Geological Survey,

"Evapotranspiration and the Water Cycle," https://water.usgs.gov/edu/watercycleevapotranspiration.html.

3. CLIMATE DYNAMICS: NATURAL VARIATIONS

1. Oxygen has two isotopes, O_{16} and O_{18}. With a temperature increase in water (H_2O), the lighter isotope preferentially evaporates, so the ratio of the two isotopes is a robust measure of water temperature at present and well into the past.
2. Earth's shape is an oblate spheroid, having a diameter at the equator 43 kilometers greater than the diameter measured pole to pole. This small asymmetry is enough to allow the moon's gravitational attraction to have a stabilizing effect on Earth's orbital motions.
3. In some branches of science, especially electrical engineering, this may be called "feed-forward."
4. There is also an effect caused by the surface roughness of the ocean; CO_2 gas exchange is promoted in rougher ocean conditions.
5. Normal conditions are sometimes referred to as "neutral conditions."
6. The thermocline may be absent at very high latitudes.
7. Bjerknes also established the notion of a weather front based on observations of troop movements during battle in World War I.
8. Center for Climate Prediction Merged Analysis of Precipitation, https://www.cpc.ncep.noaa.gov/products/global_precip/html/wpage.cmap.shtml.

4. CLIMATE IN THE FUTURE

1. The use of the term *fat tail* to describe this distribution does not originate with the author.

5. EARTH'S RESPONSES TO CLIMATE CHANGE

1. The graphs in figure 5.1 are from the Intergovernmental Panel on Climate Change Fifth assessment report UNEP 2017 and have had that correction made.
2. A knot is 1 nautical mile per hour or 1 minute of latitude per hour. A nautical mile is 1.15 miles, or 1.852 kilometers.

3. A few other conditions are also needed, one of which is that the vertical temperature gradient needs to be such that the atmosphere can support convective motion.

4. Thomas R. Knutson, John Mcbride, Johnny C. L. Chan, Kerry Andrew Emanuel, Greg Holland, Christopher W. Landsea, Isaac Held, James P. Kossin, A. K. Srivastava, and Masato Sugi, "Tropical Cyclones and Climate Change," *Nature Geoscience* 3, no. 3 (2010): 157–63.

5. GFDL summary report. https://www.gfdl.noaa.gov/global-warming -and-hurricanes/.

6. Knutson et al., "Tropical Cyclones and Climate Change."

7. Knutson et al., "Tropical Cyclones and Climate Change."

6. CLIMATE AND SUSTAINABLE DEVELOPMENT

1. Paul Collier, *The Bottom Billion: Why the Poorest Countries Are Failing and What Can Be Done About It* (New York: Oxford University Press, 2007).

2. Data files for Setting the Record Straight, archived online at https:// web.archive.org/web/20080720120636/http://www.earth-policy .org/Updates/2006/Update56_data.htm.

FURTHER READING

Bender, Morris A., Thomas R. Knutson, Robert E. Tuleya, Joseph J. Sirutis, Gabriel A. Vecchi, Stephen T. Garner, and Isaac M. Held. "Modeled Impact of Anthropogenic Warming on the Frequency of Intense Atlantic Hurricanes." *Science* 327, no. 5964 (January 2010): 454–58.

Cane, Mark A. "The Evolution of El Niño, Past and Future." *Earth and Planetary Science Letters* 230, no. 3–4 (2004): 227–40.

Cao, Long, and Ken Caldeira. "Atmospheric Carbon Dioxide Removal: Long-Term Consequences and Commitment." *Environmental Research Letters* 5, no. 2 (2010): 1–6.

Church, John A., Philip L. Woodworth, Thorkild Aarup, and W. Stanley Wilson, eds. *Understanding Sea-Level Rise and Variability.* Hoboken, N.J.: Wiley Blackwell, 2010.

Cullen, Heidi M., and Peter B. deMenocal. "North Atlantic Influence on Tigris-Euphrates." *International Journal of Climatology* 20, no. 8 (2000): 853–63.

Giannini, Alessandra, Michela Biasutti, and Michel M. Verstraete. "A Climate Model-based Review of Drought in the Sahel: Desertification, the Re-greening, and Climate Change." *Global Planetary Change* 64, no. 3–4 (2008): 119–28.

Goddard, Lisa, and Maxx Dilley. "El Niño: Catastrophe or Opportunity." *Journal of Climate* 18 (2004): 651–65.

Hoerling, Martin, J. W. Hurrell, Jon Eischeid, and Adam Phillips. "Detection and Attribution of Twentieth-Century Northern and Southern African Rainfall Change." *Journal of Climate* 19, no. 16 (2006): 3989–4008.

Hurrell, J. W. "Climate: North Atlantic and Arctic Oscillation (NAO/AO)." In *Encyclopaedia of Atmospheric Sciences*, ed. James Holton, John Pyle, and Judith Curry. Cambridge, Mass.: Academic Press, 2002.

Intergovernmental Panel on Climate Change. *Contribution of Working Group I to the Fourth Assessment Report of the Intergovernmental Panel on Climate Change*. Ed. Susan Solomon, Dahe Qin, Martin Manning, Melinda Marquis, Kristen Averyt, Melinda M. B. Tignor, Henry LeRoy Miller, and Zhenlin Chen. Cambridge: Cambridge University Press, 2007.

Knutson, Thomas R., John L. McBride, Johnny Chan, Kerry Emanuel, Greg Holland, Chris Landsea, Isaac Held, James P. Kossin, A. K. Srivastrava, and Masato Sugi. "Tropical Cyclones and Climate Change." *Nature Geoscience* 3, no. 3 (2010): 157–63.

Kump, Lee R., James F. Kasting, and Robert G. Crane. *The Earth System*. Upper Saddle River, N.J.: Pearson Prentice Hall, 1996.

Lutgens, Frederick K., and Edward J. Tarbuck. *The Atmosphere: An Introduction to Meteorology*. 9th ed. Upper Saddle River, N.J.: Prentice Hall, 2004.

Marshall, John, and R. Alan Plumb. *Atmosphere, Ocean, and Climate Dynamics: An Introductory Text*. Amsterdam: Elsevier, 2008.

McWilliams, J. C. "Modeling the Oceanic General Circulation." *Annual Review of Fluid Mechanics* 28, no. 1 (2003): 215–48.

Meinshausen, Malte, Nicolai Meinshausen, William Hare, Sarah C. B. Raper, Katja Frieler, Reto Knutti, David J. Frame, and Myles R. Allen. "Greenhouse-Gas Emission Targets for Limiting Global Warming to 2°C." *Nature* 458, no. 7242 (2009): 1158–62.

Philander, S. George. *Is the Temperature Rising? The Uncertain Science of Global Warming*. Princeton, N.J.: Princeton University Press, 1998.

Sarachik, Edward S., and Mark A. Cane. *The El Niño-Southern Oscillation Phenomenon*. Cambridge: Cambridge University Press, 2010.

Schneider, Tapio. "The General Circulation of the Atmosphere." *Annual Review of Earth and Planetary Science* 34 (2006): 655–88.

United Nations. *World Population to 2300*. Department of Economic and Social Affairs Report ST/ESA/SER.A/236. New York: United Nations, 2004.

Visbeck, Martin H., James W. Hurrell, Lorenzo Polvani, and Heidi M. Cullen. "The North Atlantic Oscillation: Past, Present, and Future." *Proceedings of the National Academy of Sciences of the United States of America* 98, no 23 (2001): 12876–77.

Zeng, Ning. "Drought in the Sahel." *Science* 302, no. 5647 (November 2003): 999–1000.

INDEX

Page numbers in *italics* represent figures or tables.